AF254417

Female Reproductive System & Herbal Healing Vs. Prescription Drugs and Their Side Effects

*Complete Illustrated, Herbal Remedies,
Prescription Drugs & their Side Effects,
Ayurveda, Kama Sutra, Observed Female Cases,
and Herbal Formulas for Holistic Health*

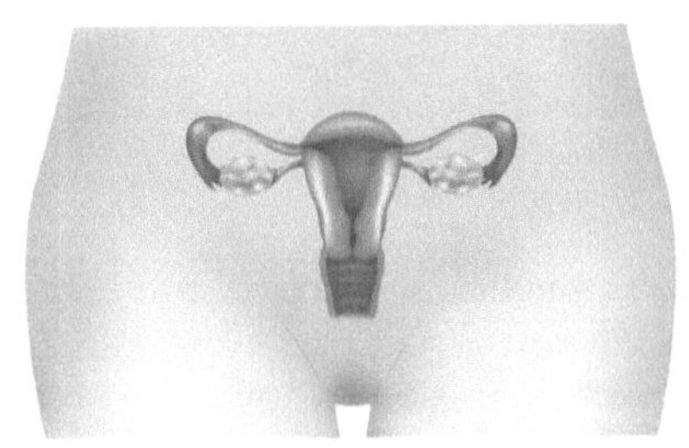

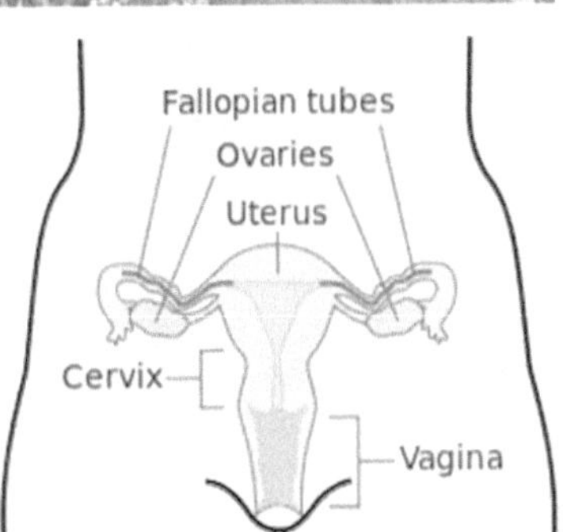

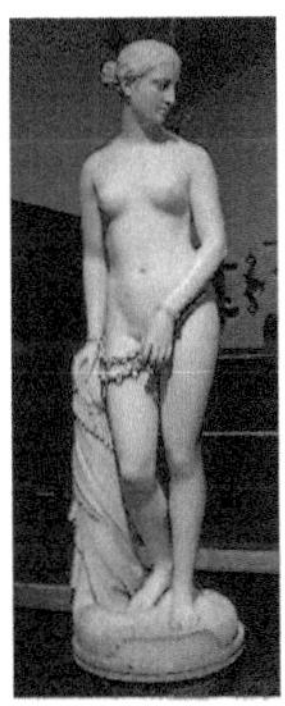

Chela Ram Bathija, MH, RH (AHG)

ISBN: 978-1-63732-252-9   (Paperback Edition)
ISBN: 978-1-63732-253-6   (Hardcover Edition)
ISBN: 978-1-63732-251-2   (E-book Edition)

Book Ordering Information

Phone Number: 315 288-7939 ext. 1000 or 347-901-4920
Email: info@globalsummithouse.com
Global Summit House
www.globalsummithouse.com

Printed in the United States of America

CONTENTS

To my wife, Rajkumari,
for her love and support to make this book possible
To my Children and their spouses who gave me all the
moral support in fulfilling the desire for completing this
book
Asha-Sanjay, Sunil-Priyamika, Babita-Dabe, and
Sheebani-Manpreet
&
My lovely grandchildren
Sheevani, Anisha, Sahil, Annika, Serina, Suhani,
Jasmeen, Armaan and Aleena

Before I became an Herbalist, I was a medical doctor with a MD degree. As a post graduate in Pediatrics I earned D.C.H (Diploma in Child Health). I did further training in United States in Neuroscience and Public Health. I have practiced for more than thirty years seeing the patients & client's of all ethnicity and ages in Asian countries, as well as in the Unites States.

Qualifications:
MD, Kabul, Afghanistan
DCH, New Delhi, India
Residency: Neuroscience and Behavioral Psychology, Louisiana State University
Certificate: Public Health, Health Promotion, George Washington University
MH (Master Herbalist), USA
RH (Registered Herbalist), USA

FOREWORD

I recommend Dr. Bathija's book on the female reproductive system. He is an accomplished herbalist and is sharing ancient wisdom blended with modern science. He expands our knowledge of the female reproductive system with needed information and illustrations. I look forward to more of his books with anticipation.

Bishop Dr. Truman Berst
Master Herbalist for over 50 years

Dr. Truman Berst is the founder of Health & Herbs Company. I have gotten the chance to know Dr. Truman Berst personally for the last 7 years, taking effect after I started my herbal practice. I have consulted many difficult cases with his guidance, which has helped many of my clients.

FOREWORD

This book is collaborative result of knowledge from Dr. Chela R. Bathija who is compassionate and very well learned healthcare provider in the field of medicine and alternative medicine. A number of prescription drugs have been reported to have negative effects hence I think author's book "female reproductive system, herbal healing vs. prescription drugs and their side effects" indeed is a very well written topic. I wish Dr. Chela R. Bathija my best of luck and many congratulations on giving this topic the needed attention.

Dr. Sona R. Lund
26 August, 2013

FOREWORD

It is my privilege to share *The Female Reproductive System and Herbal Healing versus Prescription Drugs and their Side Effects* by Dr. Chela Ram Bathija. Dr. CRB has combined up-to date scientific knowledge with a comprehensive approach to total health. There are multiple dimensions to maternal and child health. Mothers and children are vulnerable to pathologic and environmental hazards. The healthier the majority of people are, the better and brighter the future of the community will be on a global health level.

The World Health Organization defines health WHO "World Health Organization" as "physical, mental, psychological, environmental, emotional, and economic well-being and quality of life." Growing health costs and expensive pharmaceutical products with major side effects require new, effective, affordable approaches.

Herbal treatments, health education, nutrition, religious healings, and immunizations etc., are all addressed in Dr. CRB's book. *The Female Reproductive System and Herbal Healing versus Prescription Drugs and their Side Effects* covers preventive and curative aspects with a holistic approach.

Dr. Chela Ram Bathija was introduced to me through a mutual friend. He had just graduated from high school and had a desire to pursue a career in medicine. It was my pleasure to be one of his mentors while he attended medical school.

While I was teaching and serving as vice dean in the College/ Faculty of Medicine, Kabul University, Kabul, Afghanistan, he was one of my most capable students. After graduation, he served as a pediatrician at the Indira Gandhi's Children Hospital, Ministry of Public Health (MOPH) in Kabul.

His desire to be a physician was well founded. In his early years Dr. Bathija was very enthusiastic about becoming a medical doctor. He was brought up in a culture that introduced many healers of homeopathic, Ayurvedic, Greco-Roman, and Indo-Chinese healing practices. Herbal healing is the backbone of many of these practices. CRB is a naturally born herbalist.

On the teaching staff at the Department of Community of Medicine and Family Practice at, Howard University, I have found Dr. CRB's overall approach to be crucial, significant, affordable, effective, and free of side effects.

The current health system cannot offer curative aspects of health for all; therefore, herbal healing is the best strategy to utilize. Herbal medicine reaches those who cannot afford conventional medicine.

Traditional medicine is the most appropriate method to cover social physiology, social anatomy, social diagnosis, and social treatment. Initially, this approach is valuable and significantly applicable since it covers all ages, sexes, and stages of human growth and development. It also reduces mortality and morbidity and increases life spans while providing safe, healthy environments.

M. Z. Bashardost, MD
Public and Community Health Specialist

FOREWORD

Sometimes as an educator you get the opportunity to work with a student that just shines. A student who offers inspiration, observations and a passion to the work that makes it clear that the herbal world has a gifted soul. Working as Chela's teacher was an honor. He not only believed in the work and the power of herbal medicine, but he challenged himself to find answers and to mesh it all together with all of the other knowledge he already had in reference to health, healing and herbalism. I am so excited to introduce this body of work and honored to have worked with such a fine herbalist.

Demetria Clark, M.H.

NOTE TO THE READER

This book is not intended to replace the services of a physician. It is not, meant to encourage diagnosis and treatment of illness, disease, or any medical problem. The author's duty is to acquaint you with the best alternative and holistic approaches and safer remedies for better health.

It is the reader's choice to follow or not follow the recommendations presented in this book. If you are under a physician's care for any condition, he or she can advise you whether the recommendations presented in this book are suitable for you. For pregnant women, though there are many herbal products that are safe to take, it is always advisable to consult with your primary care provider before you take any herbal products.

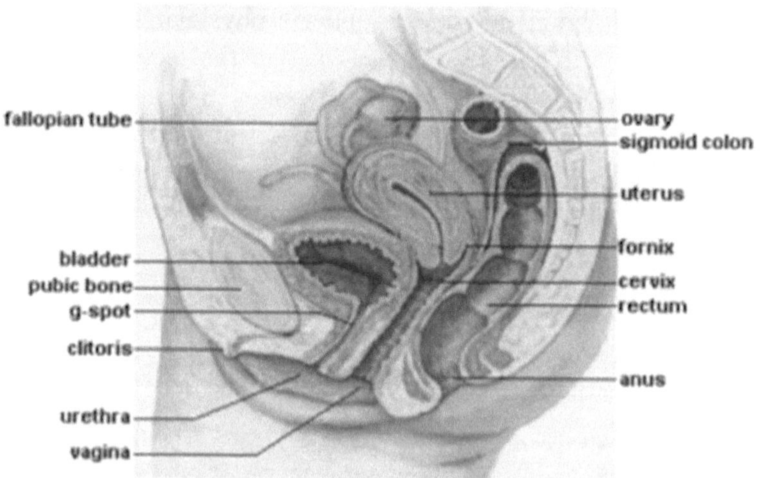

female genitalia side view

MISSION

Healing in a Holistic Way

My mission is to cure each of my clients with proper care and to educate them on how to properly take care of themselves.

Since we are living in the twenty-first century, I am proud to take this opportunity to speak about building wealth through promoting health. We can claim this idea through natural health and wellness revolution, and we would like to keep our families and ourselves healthy in natural ways.

Healing in a holistic approach will be the medicine of the future. It will include Ayurveda, natural medicine, good nutrition, herbs, and various mind-body therapies. Holistic approaches will also include yoga and meditation, which are ways to nourish the body, pamper the soul, and utilize the gifts of nature. As a nutritional consultant, I have the important task of educating others about nourishing their bodies for optimal health.

Allopathic medicine is becoming costly and has a tendency to cause side effects. The demand for herbal medicine is growing due to the increased popularity of herbs and herbal medicines. People of all ages who are suffering from various diseases are showing interest in herbs and herbal medicine, and they are using them as supplements for optimal health.

During the three decades of my medical career, I have dealt with people who have suffered from physical and emotional disorders, such as obesity, heart disease, stroke, cancer, HIV/AIDS, PTSD, insomnia, anxiety, depression, and sexual impotence. I have noticed that the whole body (body-mind- soul) is taken into consideration while using a holistic approach, whereas conventional medicine only treats the symptoms. It enhances innate healing power by supplementing the necessary herbal products to optimize immunity levels and resist environmental stresses. This is especially true in youngsters and the elderly.

In my seven years of herbal practice, I have consulted about 900 out of 1475 (out of total 1475 cases that I have consulted, 900 cases belonged to female sex) teenagers, adults, and menopausal female cases. A majority of those who have come to me have had strong beliefs about herbal remedies; this makes me feel optimistic toward my holistic approach for dealing with their health and well-being issues.

Most of the household burdens, which include earning, household chores, and raising children, are the responsibility of females. They pay minimum attention to their health. My duty as a healer is to address these concerns in my presentations and lectures.

Conventional or allopathic medicine has created a dilemma for many women because of its wide range of services. Many women have become disillusioned with conventional (allopathic) medicine. For example, patients may be sent to a gynecologist for menstrual problems, a gastroenterologist for acid reflux disease, an endocrinologist for hormonal problems, and a neurologist for headaches and migraine. The bottom line is that no one takes an overall view of that patient; the common factor in these conditions is stress, which is overlooked in many cases. This is where a consultation with a naturalist or alternative medicine therapist who looks at the whole picture can add the beauty in an individual's life.

Many women are reluctant to take drugs in any form, especially in the long term, because of potential side effects and because the drugs often address the symptoms rather than the main cause of a problem. The holistic approach teaches clients to recognize that there is more to health than the absence of illness and it teaches how to take positive steps toward making lifestyle changes, following healthier diets, exercising, and learning how to rest and relax.

I am very excited to introduce *The Female Reproductive System and Herbal Healing versus Prescription Drugs and their Side Effects* as part of an alternative approach that utilizes yoga, meditation, and acupuncture.

It's the perfect time to learn more about this exciting and rewarding field. You can enhance your life—and the lives of those around you.

AUTOBIOGRAPHY

Born in Afghanistan in 1946, after completing high school, I became an Aeronautic Engineer. My family was not happy with my decision, so I chose a different profession & decided to pursue a career in medicine. I studied medicine and graduated from Kabul University with M.D. degree and later earned a Diploma in Child Health (D.C.H.) in Pediatrics from All India Institute of Medical Sciences, New Delhi, India. I came to the United States for the sake of a better future and to continue my profession in United States Health Care System. After passing the foreign medical graduate exams I received basic and advanced training in Neurosciences and Psychiatry in Louisiana State University and studied Public Health, a course on Health Promotion in George Washington University.

Belonging to a family who loved to grow and nurture flowers and herbal plants, I developed this pure nourishing and healing nature in myself. I eventually became interested in learning more about Herbal Medicine. Primarily, I learned this field from my father and my wife. I earned the Master's in Herbal Health (M.H). After five years of my herbal practice and consulting about 1250 clients, each with different health ailments, I took the American Herbalists Guild test, and was honored to become the Registered Herbalist, R.H. (AHG) in the United States.

As per to date I have consulted 1,575 plus client's. I also offer consultations over the phone with my clients who reside in different states and overseas. I also speak, Persian, Pashtu, Urdu, Hindi and Spanish. About 80% of my regular clients consist of Latinos. Besides providing herbal remedies, I also provide Massage and Laser Therapy for the pain reduction and consultation on psychological and personal issues. In my free time, I study about Ayurvedic healing system and writing about Herbalism.

ACKNOWLEDGMENTS

The author wishes to thank the following people for their thoughtful advice in running my herbal practice and granting the permission for the use of illustrations and herbal formulations in my book:

Demetria Clark, MH: My instructor in herbal teaching, she has shaped my professional life.

Gail Faith Edwards: Instructor and author, she is a well-known herbalist who loves plants, gardening, and more. I am privileged for the permission to use her illustrations of herbs in this book.

Truman Berst, MH: Herbalist, author, consultant, and owner of Health and Herbs Co, he has assisted me with herbal consultations on difficult cases and provided me the best herbal products for my clients during the past seven years.

Natura-Genics: Manufacturers and distributors, they always help me provide the best herbal products.

Asha Kakkar: An accountant and financial advisor, she has helped me on every single occasion to establish my health-related and personal business. I am very impressed by her determination and sound judgment in completing my projects without hesitation.

Sunil Bathija: He provides full technical support in maintaining Internet assistance for completing my projects. Sunil has previously worked at the Medical College of Georgia, a teaching hospital and 400+ bed facility, as clinical/pharmacy analyst. He has used his computer knowledge to analyze drug interactions and their side effects prior to their use on patients. He has also supervised technicians and pharmacists in the use and programming of new hardware and software.

Teresa Rivero: A medical assistant, she has extensive medical knowledge of health ailments in children and women. She has assisted me by introducing the cultural, social, and lifestyles issues related to Latino populations who have come to my herbal practice in the past seven years.

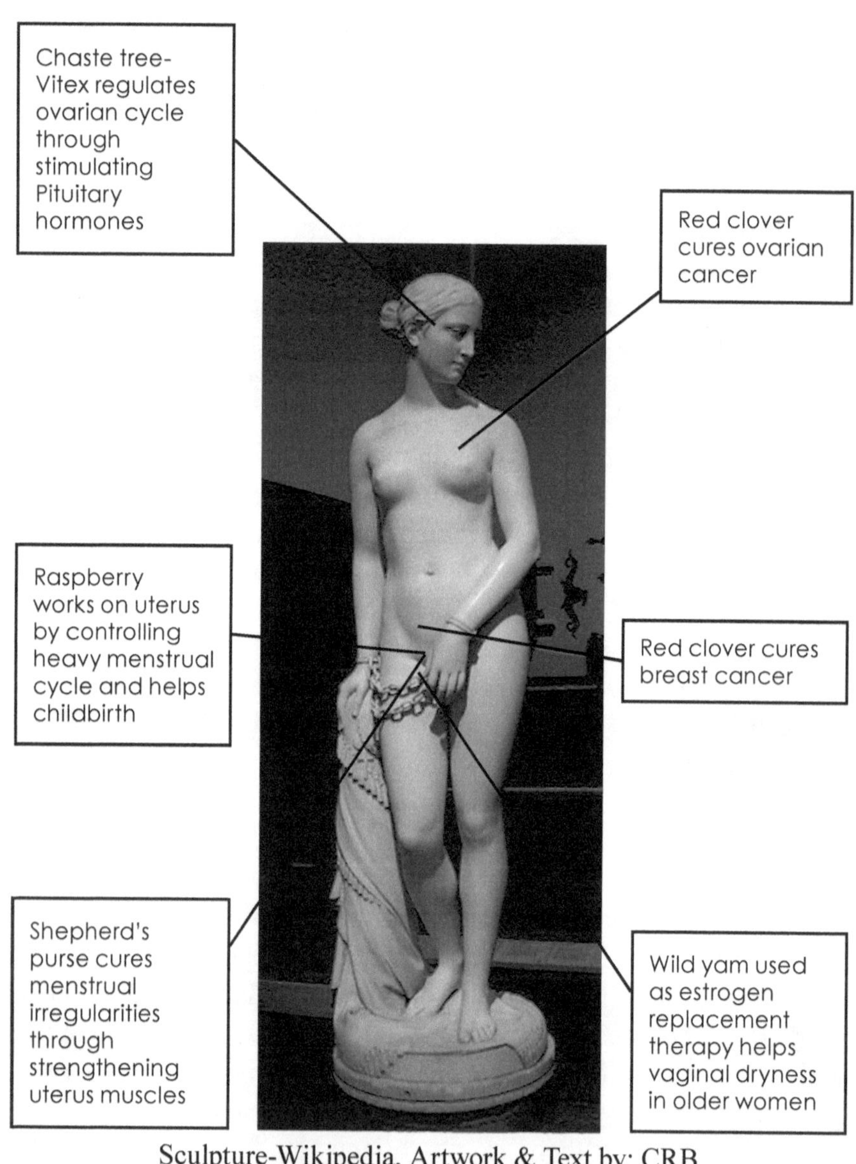

Sculpture-Wikipedia, Artwork & Text by: CRB

This is a few examples of many herbs represented in this book that have many functions curing different organs. Above, I have represented just a few herbs, showing where they can help.

HOW TO USE THIS BOOK

This book can be read in several ways. It can be primarily used to become familiar with herbalism. It can be used as a textbook for finding out about holistic treatments for specific ailments and pathology of female genitalia or as a reference book about the indication of Ayurvedic and Western herbalism.

There are nine main sections in this book. The first part, an introduction to *Female Anatomy and Physiology*, gives you an overall picture of a woman's genital structure and how the cyclical secretions of related hormones regulate the reproductive organs. The second part, *Ayurveda*, briefly and concisely introduces the traditional Indian way of healing and herbalism, which has existed for more than five thousand years. It also introduces the constitution (*prakriti*) of the human being, the three basic principles (doshas) of the human nature, the diseased states (imbalances), and the healthy states (harmony).

The third part, the *Teenage/Puberty Stage*, covers almost every single disease and discusses the unsafe lifestyles of this group— and their management with herbal remedies, nutrition, diet, and healthy lifestyle. The fourth part, *Pregnancy and Childbirth/ Childbearing Age*, offers a way of living and balancing the physical and emotional states of childbearing for working mothers.

The fifth part, *Menopause*, highlights the normal changes that women go through because of hormones that cause emotional imbalances and suffering. A proper diet and herbal supplements can alleviate the suffering and overcome any anxiety created in this phase. The sixth part, *Impotence*, deals with common female issues in female life and explains how impotence can be managed with the nutrition, herbs, and counseling. Special attention is given to counseling, focusing on social, financial, and marital conflicts.

If sexual impotence is left untreated for too long, it can be frustrating. It is a reversible phenomenon if proper management is considered through herbs, counseling, and diet; it can help a person or a couple overcome this issue. The seventh part, *Cancer,* describes the common types of cancers, the risk factors, and how to remedy and manage the predisposing factors in order to have vibrant health.

The eighth part, *Diet, Nutrition, and Vitamin,* uses the food classification pyramid system to introduce a thorough and detailed description of nutrition, diet, and essential elements and vitamins. Finally, the ninth part, *Materia Medica* , a brief classification of Ayurvedic and Western herbalism, gives useful information about the herbs used in female reproductive health and the treatment of specific conditions and problems.

Read chapter 2 to gain a basic understanding about your own *dosha* before reading the rest of the book.

CHAPTER 1

The Anatomy and Physiology of the Female Reproductive System

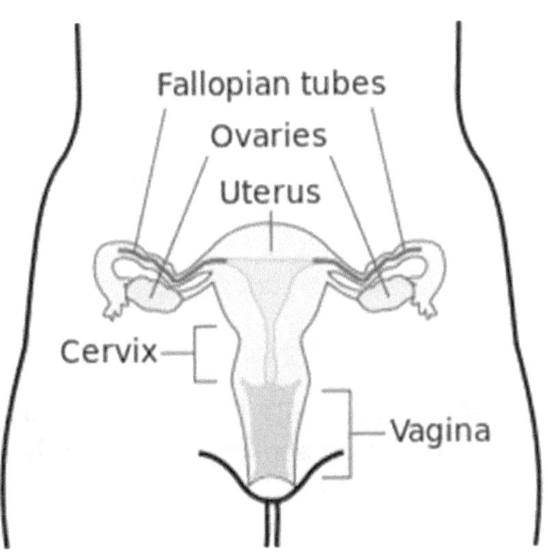

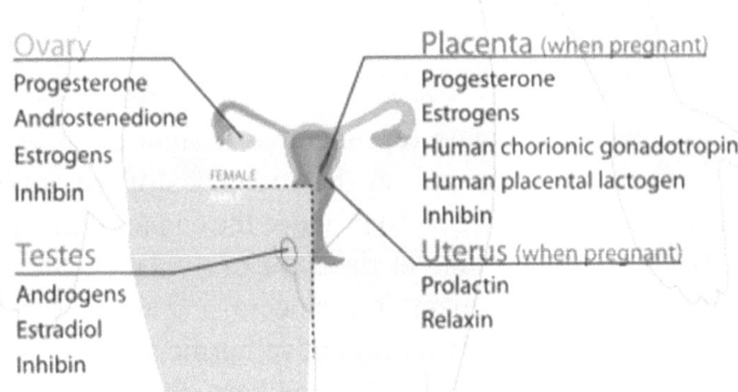

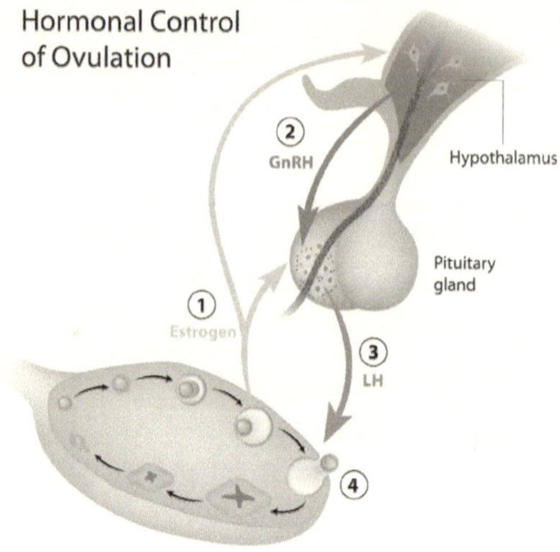

Female Organs

Uterus

It is a hollow space or a cavity in the female pelvic region. It is located in the lower part of the abdomen, between the large intestine and the urinary bladder. The primary functions of the uterus are preparation for the fertilization of the fetus, nourishment of the fetus, and taking care of the fetus during the nine months of pregnancy.

Ovaries

The ova and the female reproductive hormones are produced. The two ovaries are situated on the two sides of the uterus. The main function of the ovaries is to produce the egg (ovum) once every month. If sperm is available at the time of ovulation, the egg can unite with the sperm, fertilization can occur, and a fetus is produced. Also ovaries secrete female reproductive hormones that play a very important role in a woman's life from puberty until menopause.

Fallopian Tubes

These ducts in the female abdomen transport the egg from the ovary to the uterus. They connect the ovaries with the uterus. When the eggs are released from the ovary, they are fertilized in the first part of the fallopian tube after combining with the sperm. After the sperm and eggs are joined, they are sent to the uterus for further development and growth. The joining of the sperm and egg is called ovulation.

Vagina

The external part of the female genital tract consists of the cervix, which is attached to the uterus for expelling the fetus at the time of delivery, labia minora, labia majora, and the clitoris.

Breasts

Reproduction, lactation, and lifestyle play a major role in healthy breasts. Increase in estrogen level protects the heart and bones from aging but cause breast cancer. Healthy breasts are important for moms as well as infant for breastfeeding.

The glands that supply milk to the infant are located in upper thorax, one on each side.

The entire physiology of the female reproductive system is dependent upon the hormones secreted by the ovaries (estrogen and progesterone), which are controlled by the pituitary gland. These hormones carry out several functions in a woman's reproductive system.

The hypothalamus releases a hormone called gonadotropin releasing hormone (GnRH) . Gonadotropin releasing hormone then signals the pituitary gland to release two hormones—luteinizing hormone (LH)—to cause ovulation and follicle-stimulating hormone (FSH)—to start sexual development such as genital, breast enlargement and hair growth on genitalia.

Hormonal Secretions and their Effects on the Genital Organs

Hormones carry the following functions in a female:

- They cause the monthly menstrual flow (onset of menarche), continue for four to six days, stops the blood flow after six days, and develops the ovum (egg) once every month.
- They help in the growth and management of the female secondary sexual characteristics, such as enlargement and development of the breasts, hair growth in the armpits and on the vulva, feminization of the voice, and accumulation of fat on the buttocks and thighs.
- They help in preparing the internal uterine wall for receiving the fetus if fertilization occurs or shred the inner uterine wall if the fertilization does not take place.
- If pregnancy follows, the hormones take care of the growth of the fetus and facilitate childbirth after the maturation of the fetus is complete in the uterus.
- They help in preparing the breasts for lactation during pregnancy and cause adequate lactation after childbirth.

CHAPTER 2

Ayurveda

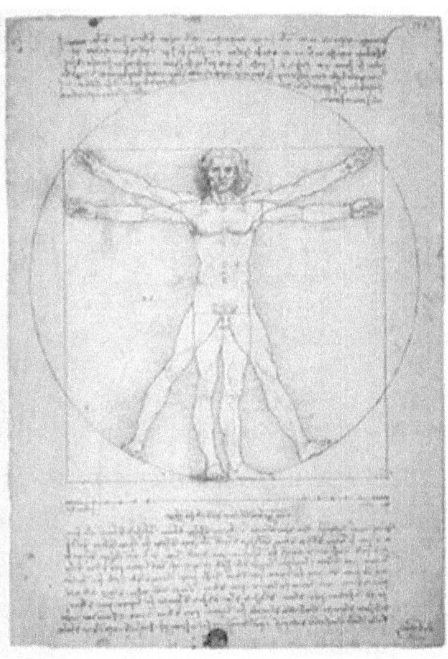

Ayurveda, the ancient knowledge of healing, has existed for more than five thousand years, depending on the data given by different sources.

Ayurveda has been written in the ancient books known as Vedas. Many writers have tried to explain it, including Swami Sadashiva Tirtha, Shyam Sunder Rao Chepur, Maharishi Mahesh Yogi, Vasant Lad, Deepak Chopra, Robert E. Svaboda, and David Frawley. These writers and philosophers believe that Ayurveda helps restore balance in the physiology, eliminates toxins and impurities, and awakens the body's natural healing mechanisms. In order to better understand the topic, it is important to explore its basic principle (*doshas*).

Thousands of years ago, pioneers, yogis, and monks were able to explore brain tumors, ovarian and kidney cysts, pancreatic and colonic cancer, hepatic lesions, and epilepsy. They were able to do this by examining through the eyes, reading pulses, or by palpation of the body organs. They did not have today's advanced technology—x-rays, sonograms, CT/MRI, and endoscopies—to discover and detect the diseases or pathologic findings.

Doshas

To understand Ayurveda, it is essential to know about the three *dosha*s (life forces). The physical and mental constitution of an individual is called *prakriti*. Vasant Lad, BA, MS, MASc, describes the three *dosha*s (*vata*, *pitta*, and *kapha*), in *Textbook of Ayurveda, Volume One: Fundamental Principles*.

Vata Dosha

Vata is associated with the elements of space and air. Any disorder of this *dosha* in any organ is called air disorder. People with dominant *vata dosha* energy are active and alert. They are cool, quiet, softhearted, and romantic. They have thin, built-up, dry skin and prominent bones. If they lose balance, they can be nervous and fearful.

Vasant Lad explains that *vata* energy is located in the colon, and these people are prone to health conditions, such as flatulence, gastro-intestinal discomfort (constipation), tics and twitches, aching joints, dry skin and hair, nervous system disorders, anxiety, and nervousness. *Vata* energy is strongest during the fall season.

Vata people are particularly sensitive to sugar, alcohol, and drugs; therefore, these people should limit the use of these substances. Avoid cold foods, especially during the fall and winter. Avoid ice cream and other cold sweets. Choose warm foods and spices and limit your intake of raw foods. Consume salads and raw vegetables in summer, preferably at lunchtime, when digestive fire (Agni) is strongest. *Vata* people live well in warm, coastal climates.

According to "Health and Ayurveda." (herbalsafari.com), most of the female reproductive system diseases are caused by *vata* (*vayu*) accumulation in the organs. Examples are dysmenorrhea, miscarriages, PMS, vaginitides, and yeast infections. Any disorder in this *dosha* make the *vayu* (air) move toward the genital tract and causes pain, irritation, and diseases that belong to *vayu*. Menstrual discharge is painful, frothy, thin, and creates sounds.

Pitta Dosha (Pit Heat)

Pitta is made of fire and water. Anyone with freckles or red, blond, or prematurely gray hair is probably *pitta*. These people have strong metabolisms and efficient digestive systems. The energy center of these people is the small intestine. These people easily maintain weight proportionate to height because of their strong metabolisms.

Pitta people are determined, passionate, and successful. They work well under pressure and handle emergencies very well. When they become unbalanced, they are quite scary; they can lash out or develop ulcers. They are very good at making money and spend it easily, but they are not so good at accumulating wealth. They are always quick to move from one passion to the next.

They tend to develop skin rashes, outbreaks, sunburns, and poison ivy reactions. Summer is their season. They are prone to different inflammations, such as conjunctivitis, colitis, sore throats, ulcers, and fevers. At menopause, *pitta* women may have trouble with hot flashes.

The optimal *pitta* diet consists of cooling foods, such as cottage cheese, mint tea, oatmeal, basmati rice, and sweet-tasting fruits. Though *pittas* often love to eat hot, spicy foods, these foods aggravate their natural fire; therefore, they need to use them sparingly.

While *vatas* may skip meals because they simply forget to eat, *pittas* are always punctual for meals and dinners. The best place for *pittas* to live is in cool climates.

The increase *pittas* cause a disturbance in this *dosha* and make *pitta* move to genital tract and produces burning, inflammation, fever, and heat. This means that heat is produced in the genital tract which causes menstrual flow to be blue, yellow, or black with excessive hot discharges and a foul odor.

Kapha Dosha (Phlegm)

The disturbance in this *dosha* is called a water disorder. This *dosha* is connected with earth and water. The people who belong to this *dosha* tend to have well-developed bodies and big bones. Their bodies turn into couch potatoes. Their hair is plentiful, dark, and curly; they have large, beautiful eyes and charming smiles. They have trouble controlling their weight. *Kaphas* are strong, calm, and forgiving.

Kaphas are wise, relaxed, and tolerant. They are grounded and readily accept changes. They have good humor and make friends easily. *Kaphas* also become attached to people and things. If they go downhill, they end up becoming greedy, selfish, and possessive, which is an obstacle to spiritual health.

Kapha energy is dominant in winter and early spring, and the diseases they acquire are colds, flues, sinusitis, and seasonal allergies. They often feel tired and retain water.

They can live long and have the advantage of a healthy lifestyle. They possess natural strength if they eat sensibly and exercise regularly. A good *kapha* diet consists of pungent, bitter, and astringent foods. They need to avoid sweets and follow a low-fat diet in order to stay healthy.

They are good at making money and saving it. They love to live in warm or moderate weather.

Excessively watery foods and liquids cause toxins to move to the female reproductive tract. This change causes them to experience slimy, cold, itchy, mild pain. The menstrual flow is pale and slimy.

Female Reproductive System

I will divide all the abnormal findings of female reproductive organ in three age groups:

- **Teenage/Puberty period**
- **Childbearing Age**
- **Menopause**

CHAPTER 3

Teenage/Puberty Period

The Normal Menstrual Cycle and the Role of Estrogen/Progesterone

A normal menstrual cycle is menstrual bleeding that occurs regularly every month. A series of changes occurs in a woman's body to prepare her for pregnancy. Although the average cycle is twenty-eight days, it can be shorter or longer. Girls usually start their menstrual cycles between the ages of eleven and fourteen. When a woman reaches thirty-nine, she gets fewer and fewer periods until periods stop and menopause starts around the age of fifty.

Consult your doctor if you have three or more very heavy menstrual periods that last longer than seven days, are bleeding between periods, or are having pelvic pain that is not from your period.

Estrogen and progesterone play the biggest roles in how the uterus changes during each cycle. Estrogen builds up the lining of the uterus, and progesterone increases after an ovary releases an egg (ovulation) at the middle of the cycle. This helps the estrogen keep the lining thick and ready for the fertilized egg. A drop in progesterone (along with estrogen) causes the lining to break down. This is when your period starts.

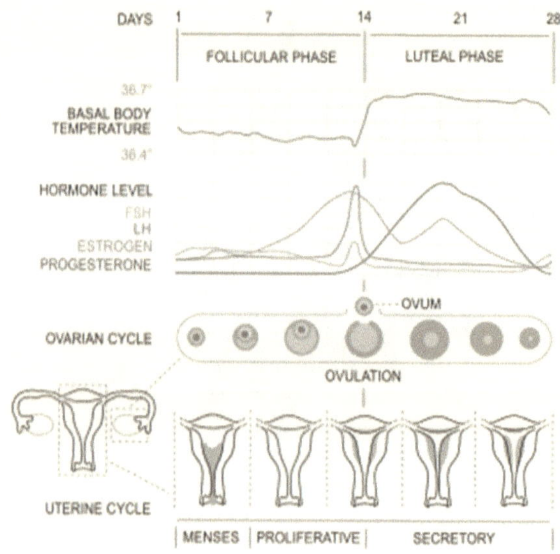

From Wikipedia, Menstrual Cycle

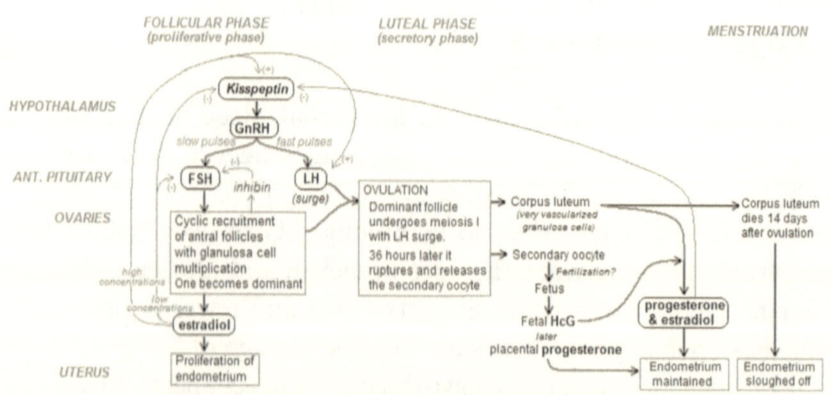

From Wikipedia, Menstrual Cycle

The hormonal disturbance can cause menstrual irregularities, which will be discussed in the next chapter. Menstrual irregularities can bear different names, such as dysmenorrheal (painful menstruation), amenorrhea (no menses), and polymenorrhea (heavy menstruation). Some females who do not have excessive bleeding during menstruation may experience premenstrual syndrome, widely known as PMS.

The major problems in this period are: Vaginities, Teenage pregnancy, and HIV/AIDS.

Other less common ailments in this period include endometriosis, ovarian cysts, and menstrual disorders, which will be discussed in the childbearing period.

Vaginitis

Vaginitis (inflammation of the vagina) is the result of bacterial or yeast infections, trichomoniasis, chlamydia, viral infections, and low estrogen levels (White and Foster 2000, 533). The symptoms of vaginitis may include vaginal discharge with a foul odor, an itching or burning sensation in the vagina, abdominal discomfort, pain during urination or intercourse, and light vaginal bleeding.

Vaginitis is a common dilemma in the teen world; it is commonly known as a sexually transmitted disease (STD). However, it is also common during the childbearing period because of unprotected sex or medical conditions, such as diabetes. STDs are caused by fungal, viral, or bacterial infections, but not all vaginitides cases are STDs.

Dr. Robert M. Giller has beautifully classified vaginitis in *Natural Treatment and Vitamin Therapies for over 100 Common Ailments*. According to Dr. Giller, non-STD irritants can cause vaginites. These include chemicals, laundry detergents, spermicides, feminine hygiene products, latex condoms, soaps, foreign bodies— such as tampons or diaphragms—left in too long, traumatic sexual activities, or physical trauma to the vagina. Vaginitis can also occur due to hormonal irregularities in postmenopausal women (atrophic vaginitis), women whose ovaries have been removed, or hormonal imbalances that cause increased vaginal discharge.

It is important to highlight some major concerns about how vaginitis occurs and how it can be treated.

Risk Factors

- menopause
- diabetes
- antibiotics
- synthetic
- underwear
- douching
- unsafe sex practices

Sexually Transmitted Diseases and Genital Lesions

Trichomoniasis, herpes (85 percent HSV2), chlamydia, gonorrhea, chancroid, primary syphilis, parasites (scabies), human papillomavirus (HPV), painless condylomas, and HIV/AIDS can cause genital lesions.

Vaginal Discharge and itching can be caused by trichomonas, bacterial vaginosis (chlamydia and gonorrhea), and candidiasis.

If you get frequent vaginal infections, think of three things:

- Your infection is mismanaged—or not adequately treated—by your health care provider.
- Your infection has progressed to an advanced stage, and your local treatment cannot clear the infection. This is common with candidiasis.
- Your significant other or spouse is infected and keeps infecting you. In this case, your sex partner should be also treated.

Who is at Risk?

- Women who participate in high-risk sexual behavior, such as multiple sex partners, anal sex, and unprotected sex.
- Women who abuse intravenous drugs
- Women with impaired immune systems

Adolescent health by WHO (World Health Organization)

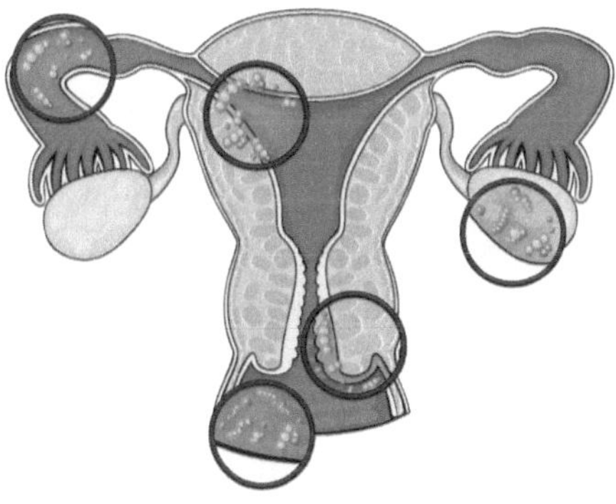

Gonorrhea

Adolescent Health

The WHO defines adolescence as a period of growth between the ages of ten through nineteen years, and they are often thought of as a healthy group. Nevertheless, many adolescents die prematurely due to accidents, suicide, violence, pregnancy-related complications and other illnesses that are preventable or treatable. Many more suffer chronic ill health and disability.

In addition, many serious diseases in adulthood have their roots in adolescence. For example, tobacco use, sexually transmitted infections (including HIV), and poor eating and exercise habits can lead to illness or premature death later in life.

Contact your doctor if you experience any of these symptoms: burning or itching in the vagina, pain during sex or urination, discharge from the vagina, sores, warts, or lumps in the genital or anal area.

Sexually Transmitted Infections

A sexually transmitted disease (STD) is an illness spread through sexual contact. These diseases can be transmitted by vaginal, anal, or oral sex.

Throughout the world, many youths have sexual intercourse and are at risk for sexually transmitted infections (STIs), including HIV, or unintended pregnancies.

The most common STDs that affect women include chlamydia, gonorrhea, herpes simplex virus (HSV), human papillomavirus (HPV), trichomoniasis, syphilis, and human immunodeficiency virus (HIV).

Premarital sexual intercourse is common and appears to be on the rise in all regions of the world. Young people everywhere reach puberty earlier and marry later than in the past. As a result, youths are sexually mature for a longer period prior to marriage.

Sexual experience varies with by region. Studies suggest that 2-11 percent of Asian women have had sexual intercourse by eighteen; 12-44 percent of Latin American women by age sixteen; and 45-52 percent of sub-Saharan African women by age nineteen. In developed countries, most young women have had sex prior to twenty: 67 percent in France, 79 percent in Great Britain, and 71 percent in the United States (Darroch et al. 2001).

Adolescent pregnancy and childbearing are associated with pregnancy complications, illegal or unsafe abortions, and death—especially in women under age fifteen (Darroch et al. 2001). Teenage pregnancy rates vary in developed countries. The ratios are 22 percent in the United States, 15 percent in Great Britain, 11 percent in Canada, 6 percent in France, and 4 percent in Sweden. It seems that childbearing pregnancy in adolescence is more common than these developed countries (Darroch et al. 2001). In other words teenage girls in America are more prone to pregnancy, then those of Canada and European countries.

Pregnant women with STDs can pass these organisms on to their children and can cause serious consequences for newborns. Pregnant women should be screened for STDs and treated immediately. Common STDs include gonorrhea, chlamydia, candidiasis, and herpes simplex.

Observed Cases

- A nineteen-year-old patient came to my office, complaining of vaginal discharge and itching for four days.
- A pregnant woman (age twenty-eight) in her second trimester—with two children (ages five and seven)—came

to my office with periodic vaginal burning and itching, which got worse during urination. She also complained of chronic nasal stuffiness, mild emotional irritability, and mood swings. She craved bread and ate some form of pasta or bread with almost every meal.

Candidacies

Genital candidacies occurs in the vagina; 75 percent of all women of childbearing age develop vaginal candidacies (hivinsite. ucsf.edu/ InSite, pg-kb-05-02-03).

Candidacies can occur in any period, but patients feel very uncomfortable during pregnancy. Fungus that is normally present in the vagina or the normal flora of the vagina is converted to pathogenic candidacies due to a change in diet, diabetes, recurrent vaginal infections, etc. A woman typically experiences itching, burning, and cheesy discharge from the vagina.

Drug Treatment

For vaginal discomfort from fungal or infectious etiology, antifungal creams such as Nystatin, Miconazole, and Clotrimazole are available. Side effects include itching, burning, and rashes.

Antibacterial creams such as clindamycin and metronidazole kill organisms that cause bacterial vaginosis. Side effects include itching, rashes, ringing in the ears, and diarrhea.

Prevention

According to Ayurvedic principles, *vata* is the *dosha* most commonly involved in vaginal yeast infections (Padove 2003).

Always practice good hygiene; limit baths; wear cotton underwear; avoid scented sanitary napkins and tampons (chemicals are used in those); wipe from front to back; never douche (it disrupts the normal flora of the vagina); consume yogurt containing active lactobacillus cultures; practice safer sex.

Diet and Nutrition

The most important step in fighting candidacies or vaginal fungal infection is modifying the diet. Start with a yeast-free, sugar-free diet and continue the diet for at least thirty days. Foods containing yeast include breads, baked goods, cheese, mushrooms, vinegar, and soy sauce.

Fermented foods, such as olives, pickles, and alcohol, sodas, dried fruits, chocolate, sweeteners, fruit juice, milk, and milk products, such as ice cream and yogurt, should be avoided.

Herbal Remedies

Garlic is an effective antifungal agent. It can be taken in its natural form or in the form of a capsule.

Lactobacillus acidophilus, the live culture found in yogurt, is beneficial in fighting Candida. Caprice acid, a naturally occurring fatty acid, is helpful for fighting candidacies.

If the patient is suffering from severe symptoms, I start them on lactobacillus acidophilus and caprylic acid. I also add antifungal garlic supplement capsules, which are available in health stores. If the client is not pregnant, tea tree oil seems to be very effective. I have used tea tree suppositories, which have 200 mg of tea tree oil. I also recommend 15 percent tea tree oil for vaginal hygiene, it should be applies externally to the infected area. Avoid any essential oils, such as grape seed extract or tea tree oil with this client to prevent a miscarriage or other complications to the fetus. Once the symptoms subside, the emotional distress and irritability will disappear.

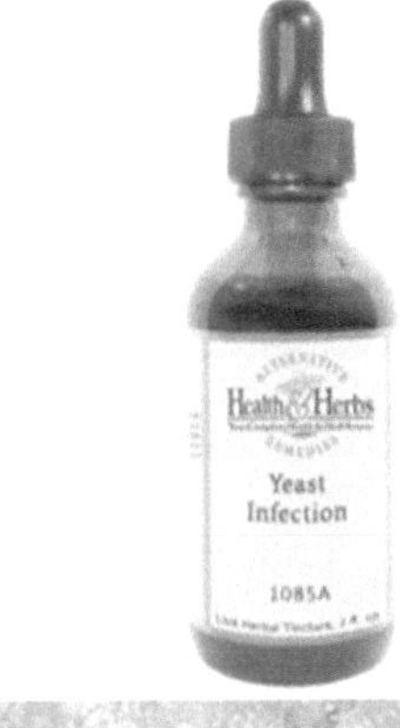

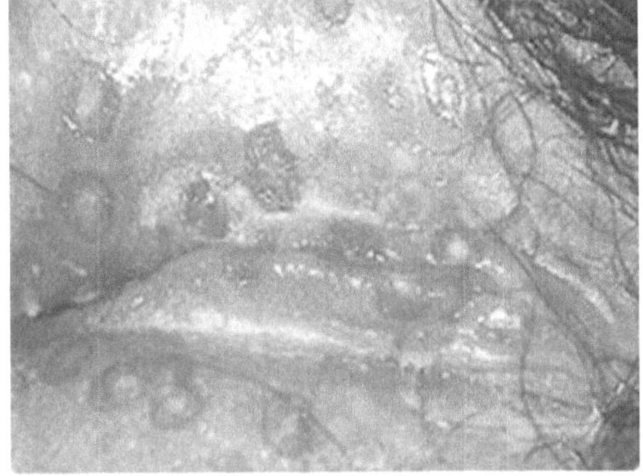

Herpes Simplex

Cases

- A twenty-one-year-old patient came to my office complaining of tender vulvar lesions. She was diagnosed with vaginal and cervical ulcerative lesions. Her inguinal nodes were tender, and her temperature was 101.2 degrees.
- A twenty-four-year-old patient came to my office after contracting herpes from sex with a male partner. She had been itching for a month.

A virus causes herpes; herpes simplex type II occurs after sexual contact with another person. It is not curable. The first sign of infection is usually genital itching. The itching is followed by painful sores, swollen lymph nodes, sore muscles, and headaches.

The outbreaks, which last from a few days to two weeks, are followed by blisters that heal in one to three weeks.

Herpes outbreaks are triggered by stress, sun exposure, weakened immunity, poor diet, surgery, skin rashes, menstruation, hormonal fluctuation, and prolonged sexual activity.

Drug Treatment

Antiviral prescription medication (Acyclovir) is available, but it does not eliminate the virus.

Side effects include excessive sweating, constipation, loss of appetite, nausea, vomiting, dizziness, headache, and confusion (White and Foster 2000, 318).

Foods and Diets that Manage Herpes

Eliminate arginine from your diet, including chocolate, walnuts, hazelnuts, peanuts, and peanut butter. Consume abundant amounts of lysine, which prevents the outbreaks. Eat plenty of turkey, chicken, fish, cottage cheese, ricotta cheese, and wheat germ. You can also take lysine in capsule form (3,000 mg lysine per day) for three months. Take zinc at the first outbreak (30-60 mg per day).

Topical Herpes Treatment

- 1/2 teaspoon St. John's wort tincture
- 1/2 teaspoon licorice root tincture
- 5 drops tea tree essential oil
- 3 drops myrrh essential oil

Combine all ingredients, shake well, and apply to herpes lesions three times a day (White and Foster 2000, 317).

Note: If tincture is made in alcohol, it stings. You can substitute alcohol with glycerates or infuse the oils of the same herbs.

Herbal remedies, such as lemon balm cream or ointment, can be applied topically to the sores four times per day. Other effective herbs, such as aloe vera, calendula, chamomile, and plantain are used to ease the symptoms. Echinacea is given for improving the immune system and should be started as soon as an outbreak begins. Continue for at least two weeks (300-400 mg capsules per day or sixty drops of tincture three times a day).

I have managed herpes with a tincture containing echinacea, goldenseal, myrrh, plantain, slippery elm, barberry, wild carrot in 20 percent alcohol and preparation by the name, VD/STD (venereal disease/sexual transmitted disease) containing black walnut, wormwood, clove, senna leaf, usnea, and male fern in 20 percent alcohol (Berst,2013).

I have suggested the above mixture with a dose of thirty drops three times a day for three to six weeks, and clients have become symptom free by following the above regimen.

Herpes Immune System Tincture

- 1/2 teaspoon echinacea
- 1/2 teaspoon St. John's wort
- 1/2 teaspoon bupleurum
- 1/2 teaspoon licorice
- 1/2 teaspoon barberry

Make the tincture from the above ingredients. Take one dropperful of the mix four to six times a day from the start until the symptoms disappear.

(White and Foster 2000, 320)

Other causes of sexually transmitted diseases:

The Duke Encyclopedia of New Medicine: Conventional and Alternative Medicine for All Ages examines sexually transmitted diseases that cause vaginitis.

Chlamydia

The greatest risk for Chlamydia infection occurs between the ages of fifteen and twenty-four. Infertility, ectopic pregnancy and pelvic inflammatory disease are major concerns if left untreated.

Symptoms of chlamydia are vaginal itching, a vaginal discharge, and painful urination, pain during sexual intercourse, lower abdominal pain, lower back pain, and bleeding between periods.

Gonorrhea

Gonorrhea is similar to chlamydia. The patient does not realize that she is infected with gonorrhea. The symptoms of gonorrhea include a vaginal discharge that can be yellowish, thick, cloudy, or bloody; a burning sensation during urination; frequent urination; pain during sexual intercourse; and bleeding between menstrual periods.

Trichomonas

Trichomonas presents with a yellow-green and sometimes frothy vaginal discharge. Symptoms of trichomoniasis include vaginal irritation, lower abdominal pain, painful urination, and discomfort during sexual intercourse.

Teenage Pregnancy

Teen pregnancy and childbearing bring substantial, considerable, and important social and economic costs through the immediate and long-term impacts on teen parents and their children.

Pregnancy and birth are significant contributors to high school dropout rates among girls. Only about 50 percent of teen mothers receive a high school diploma by twenty-two years of age, versus approximately 90 percent of women who had not given birth during adolescence (Center for Disease Control and Prevention 2011).

The children of teenage mothers are more likely to have lower school achievement and drop out of high school, have more health problems, be incarcerated at some time during adolescence, give birth as teenager, and face unemployment as young adults.

These effects remain for teen mothers and their children even after adjusting for those factors that increased the teenager's risk for pregnancy, such as growing up in poverty, having parents with low levels of education, growing up in a single-parent family, and having poor performance in school.

A teen pregnancy prevention program is one of the Center of Disease Control's top six priorities. The CDC addresses specific protective factors on the basis of knowledge, skills, beliefs, and attitudes related to teen pregnancy:

- knowledge of sexual issues, HIV, STDs as discussed previously in vaginitis section
- prevention of HIV risk
- personal values about sex and abstinence
- attitudes toward condoms
- intent to abstain from sex or limit number of partners

- communication with parents or other adults about sex, condoms, and contraception
- individual ability to avoid HIV/STD risk and risk behaviors

HIV/AIDS

HIV/AIDS (human immune deficiency virus /acquired immune deficiency syndrome) is a disease entity of the virus etiology that affects many organs in the body, depending on the time since the disease has been contacted and the person's level of immunity.

Mankind has known about HIV/AIDS for the last four decades. Its worldwide devastating health problems have endangered and killed millions of people. It is a life-threatening condition.

The HIV virus destroys the immune system by destroying specific lymphocytes in the body. Eventually, the patient experiences flu-like symptoms, swollen lymph nodes, diarrhea, and weight loss. Patients with advanced stages of AIDS get recurring infections, chills, fever, fatigue, persistent headaches, and impaired vision. It can take ten to fifteen years for an HIV-infected person to develop AIDS.

HIV is transmitted through unprotected sexual intercourse, anal intercourse, or oral contact. This happens when semen is introduced through vaginal secretions or through blood of people infected with the HIV virus. Sharing infected needles also transmits HIV. Hospital workers can be exposed to HIV by accidental needle pricks from blood transfusions, but blood is carefully screened for HIV in the United States. During pregnancy, childbirth, and breastfeeding, mothers can pass HIV to their children.

Once the virus enters the bloodstream, it invades helper T cells and reproduces itself. Eventually, the number of helper T cells circulating within the body decreases to such an extent that the body can no longer protect itself—and the immune system loses its ability to fight off infections.

In advanced stages, the patient gets central nervous system problems, such as encephalitis, meningitis, and cancerous lesions in the brain and other vital organs.

Treatments

There is no cure for HIV or AIDS. Modern medicine strives to preserve the quality of life of an infected individual and prevent the virus from replicating.

Treatment involves the use of antiretroviral drugs (often in combinations of three or more medications).

Side Effects

ART (antiretroviral therapy) causes numerous side effects and is divided into subjective manifestations and objective findings of the patient. Generally speaking, the patient may experience liver toxicity, nausea, fever, fatigue, and skin rashes. Other adverse reactions to ART are psychological changes, mood swings, agitation, attitude and behavioral problems, and so on.

In rare cases, serious medical complications include abnormal fat distribution, abnormal lipid and glucose metabolism, and bone loss.

Prevention

HIV and AIDS can be prevented by following certain guidelines:

- Before engaging in any form of sexual activity, both partners should be aware of their own and each other's HIV status.
- Intravenous drug users should seek treatment and not share needles.
- Since HIV risk increases in those who also have sexually transmitted disease, people infected with STDs should be tested and treated for HIV immediately and abstain from any sexual contact until the illness is resolved.
- Pregnant women should be prescreened for HIV infection and treated immediately if they have contracted HIV to prevent transmission to the unborn baby.
- If a blood transfusion is necessary, an HIV test should be completed immediately in emergency situations.

Meditation and Massage

This treatment is suggested for those who are suffering from extreme stress from having HIV and/or AIDS.

A combination of meditation and massage has been shown to improve the overall quality of life for some individuals with HIV/ AIDS.

There were a couple of studies done in 1979 at the University o f Massachusetts Medical Center about reducing stress. During the 1990s, Harvard University found that meditation has been linked with changes in brain structures.

Recent evidence shows that certain cortical areas are thicker in those people who meditate regularly and have found increases in blood flow in cortical regions.

It has been shown to produce a shift in brain activity away from regions associated with stress and fear responses to those that are active when a person is calm. Beneficial effects on the immune system suggest that meditation may help the body fight with the disease through increases in resistance and healing.

Herbal Treatments

Limited studies indicate that Chinese herbal compounds such as IGM-1 and SH may provide some beneficial improvement in the quality of life (*The Duke Encyclopedia of New Medicine* 2006).

Chickweed nourishes the glandular and lymphatic systems. Fresh tincture in a dose of forty drops twice daily can boost immunity and help the T cells to function optimally (Weed 1989).

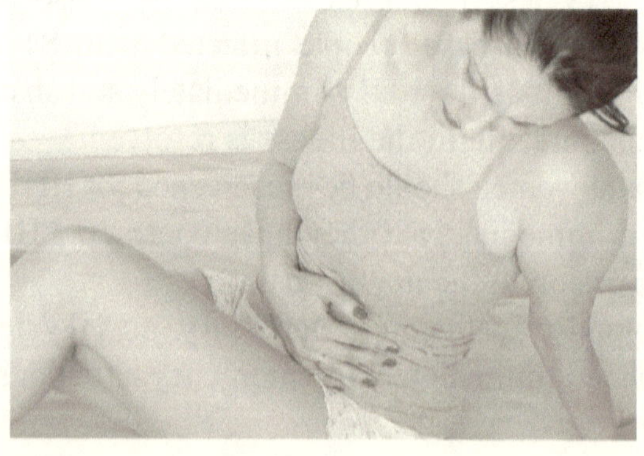

Premenstrual Syndrome

By definition, premenstrual syndrome (PMS) is a collection of psychological, physical, and emotional symptoms that occur in menstruating women in their late twenties through early forties. PMS may begin at puberty when the major changes take place in your hormone levels. PMS ends after the period begins. PMS commonly affects 40 percent of women in the United States.

Physical symptoms include fatigue, breast tenderness, swollen abdomen, oily skin that is prone to spots, fluid retention, recurrent headaches, thrush, pain at ovulation, cramping pain before or at the onset of a period, lowered libido, and food cravings (salt and sugar). Women also get flu-like symptoms, which include cold, flu, body aches, and general malaise because of decrease in immunity.

Emotional symptoms include mood swings, lack of concentration, poor coordination, poor sleep patterns, irritability, depression, anxiety, and nervousness

Psychological factors vary, but they commonly include tension, anxiety, inner responses to environmental and cultural issues, disturbance of sleep and eating patterns, attitudes toward relatives, parents, extended family, and reactions to external stimuli and childhood experiences. If the responses to these experiences are blocked, the female faces discomfort during her menstrual period.

This syndrome begins three to seven days before menstruation and ranges from mere discomfort and fatigue to debilitating cramps. Many young girls and women suffer pain and cramps due to magnesium deficiency.

Other symptoms are nausea, vomiting, mood swings, agitation, nervousness, and weight gain. These signs and symptoms are caused by the body's responses to the hormonal changes at the start of the syndrome.

Observed Cases

- An eighteen-year-old patient came to my office complaining of emotional instability, bloating, and breast tenderness. The symptoms had been so severe in the week prior to her menses that she had difficulty functioning at home and at work. She needed extreme help.

- A thirteen-year-old girl who lives with her parents had started her first period about two months earlier. She gained about fifteen pounds and developed some acne on her face. She eats pizza about twice a week with extra cheese and drinks sodas almost every day. She experiences painful menstruation for the first three days, but it subsides on the fourth or fifth day. She takes different over-the-counter painkillers.

Medications

Clonazepam (Xanax) is classified in the benzodiazepine family. It is a controlled substance that is addictive. It has been used to decrease PMS symptoms. There are many preparations available over the counter to decrease the symptoms of PMS such as, NSADS, antidepressants, diuretics, birth control pills, and antianxiety medications. All of these preparations work to some extent but have serious side effects (White and Foster 2000, 533).

Xanax causes sedation and addiction. NSADS such as ibuprofen cause gastrointestinal problems, ulcers, stomach perforations, bleeding, congestive heart failure, and reduced blood flow to the kidneys, fluid retention, heart attacks, and hypertension. Diuretics help eliminate excess water from the body, which helps with weight reduction. Common side effects are decreases in potassium levels, which cause constipation, irregular heartbeats, excessive hairiness, deeper voices, kidney/liver dysfunction, and mental confusion (White and Foster 2000, 533).

Antidepressants are the Selective Serotonin Reuptake Inhibitors (SSRIs) prescribed as PMS medications, which raise brain serotonin level and lower the severity of PMS symptoms. Potential side effects of antidepressants include weight loss or gain, loss of libido or difficulties with arousal or orgasm, insomnia, dry mouth, nausea, headache, and fatigue. These medications provide warnings that their use may lead to increased risk of suicidal thoughts (White and Foster 2000, 410).

Antianxiety medications, commonly called benzodiazepines, are prescribed for treating PMS. Common side effects include drowsiness, increased appetite, respiratory distress, amnesia/ forgetfulness, difficulty concentrating, irritability, hyperactivity, hallucinations, and suicidal thoughts.

Hormone therapy (birth control pills) is used to decrease PMS symptoms by interrupting the normal hormonal cycle and balancing the estrogen and progesterone levels in the body. Side effects of birth control pills include headache, nausea, breast tenderness or discharge, breakthrough bleeding, weight gain, emotional outbursts, migraine headaches, blood clots, and heart attacks.

Prevention

Avoid processed foods, white sugar, fried foods, coffee, and sodas. Avoid modern soy products, which tend to upset female hormones.

A week and a half before the cycle is beginning, increase calcium levels by eating kefir, yogurt, and dark green, leafy vegetables.

To prevent cramps, you can increase calcium intake ten days prior to your period. Eat calcium- rich foods and herbs, such as seaweed, sesame seeds, nuts, nettle, parsley, oat straw, and horsetail (Giller and Mathews 1994, 5).

Ayurveda

The principles of Ayurveda deal with the balance of the mind and body based on the whole body system rather than dealing with only symptoms. When *vata* is out of balance, anxiety and other nervous disorders may be present. Digestive problems, constipation, menstrual cramps, and premenstrual pains all disturb the body's systems. In order to bring this *dosha* (*vata*) into balance, it is better to avoid consuming junk food, cold drinks, and sour foods. These symptoms are aggravated by wind, caffeine, irregular meals, cold, and dry weather. Since the colon is the energy center in *vata dosha*, always keep your colon clean and hygienic.

Though some symptoms from *pitta* (agitation, hot temper, and spotting between periods) and *kapha dosha* (crying and swollen breasts) occur, *vata dosha* is important for balancing the other *dosha*s.

Complementary and Alternative Treatments

The Mayo Clinic's *Integrating the Best of Natural Therapies with Conventional Medicine, Second Edition* claims the following products can soothe the symptoms of premenstrual syndrome:

- calcium (1,000 to 1,200 mg daily)
- magnesium (400 mg daily)
- vitamin B6 (50-100 mg daily)
- vitamin E (400 IU daily)
- Chaste berry-May reduce breast pain, swelling, constipation, irritability, depression, anger, and headache. Do not use chaste berry if you are pregnant or breastfeeding.
- Ginkgo-May decrease fluid retention associated with PMS.
- Natural progesterone cream, derived from wild yam and soybeans, relaxation therapy, progressive muscle relaxation, or deep breathing exercises—can help reduce headaches, anxiety, and insomnia.

Herbs Used to Ease the Tension from Premenstrual Syndrome

PMS Tea

- 1 teaspoon *Vitex* berries
- 1 teaspoon wild yam root
- 1/2 teaspoon burdock root
- 1/2 teaspoon dandelion root
- 1/2 teaspoon feverfew leaf
- 1 teaspoon orange peel, licorice root, or stevia

Boil all herbs in four cups of water, turn off the heat, and let steep for twenty minutes. Stain the herbs. Drink two cups daily as needed (White and Foster 2000, 414)

Cramp-Relieving Tea

- 1/2 teaspoon dried motherwort herb (Gail)
- 1 teaspoon dried cramp bark
- 1/2 teaspoon dried chamomile flowers
- 1/2 teaspoon dried wild yam root
- 1/2 teaspoon dried California poppy, whole plant or root
- 1/2 teaspoon dried valerian root
- 1/2 teaspoon dried skullcap herb licorice or stevia
- 4 cups of water

Combine all in a pan, boil, lower heat, and simmer for five minutes. Stain and discard the herbs. Drink one cup as often as needed. Cramp bark and pasque flower might also be used. If water retention occurs, dandelion can be added as a mixture (White and Foster 2000, 317).

If the menstrual period is too heavy, use this formula:

- 1 teaspoon shepherd's purse tincture
- 1 teaspoon yarrow tincture
- 1/2 teaspoon red raspberry leaf tincture
- 1/2 teaspoon *Vitex* tincture

Combine all the ingredients mentioned above in a dark jar and then seal the jar tightly, leaving the jar untouched for three to four weeks. On the fourth week, take three full droppers every fifteen to thirty minutes for very heavy bleeding or two droppers every hour for moderately heavy bleeding (White and Foster 2000, 410).

Shepherd's purse tincture should be purchased from the fresh herb; it loses its strength if dried.

Vitex

If the use of contraceptive pills is the cause of the disturbance of the menstrual cycle (disturbing the normal physiology of the menstrual cycle-release of estrogen and progesterone), then stop taking the contraceptive pills and give black cohosh, chaste tree berry, licorice, and motherwort in equal parts, three times for the first two weeks after coming off the pills, twice daily for the third week, and once a day for the fourth week.

Period-Regulating Herbs

This formula, which is used as an extract, helps regulate hormones and improve moods. Use it to normalize periods and ease premenstrual syndrome.

- 1/4 teaspoon *Vitex* tincture
- 3/4 teaspoon black cohosh tincture
- 1/2 teaspoon motherwort tincture
- 1 teaspoon boldo or burdock root tincture

- 1 teaspoon St. John's wort tincture
- 1/2 teaspoon California poppy tincture
- Tincture of licorice or stevia to sweeten (optional)

Blend the tinctures together. Take one teaspoon in the morning or evening at least thirty minutes before and thirty minutes after meals. Take the preparation for at least four months (White and Foster 2000, 415).

The herbs that are given in this condition are angelica, calamus root, and castor oil packs.

Vitex is given to increase progesterone production, and dong quai is for estrogen production. Nutrients should also be taken. Sedative herbs such as Skullcap and Valerian (to be taken in equal parts) are very effective.

Summary

In the section of female reproductive system—PMS, you learned about these herbs.

Nettles, Skullcap, Valerian, Cramp Bark, Pasque Flower, Dandelion, Black Cohosh, Chaste tree Berry, Licorice, Motherwort. Here, Dandelion works as diuretics or water pills. It will not decrease the potassium level in the body contrary to water pills or diuretics.

CHAPTER 4

Pregnancy and Childbirth/Childbearing Age

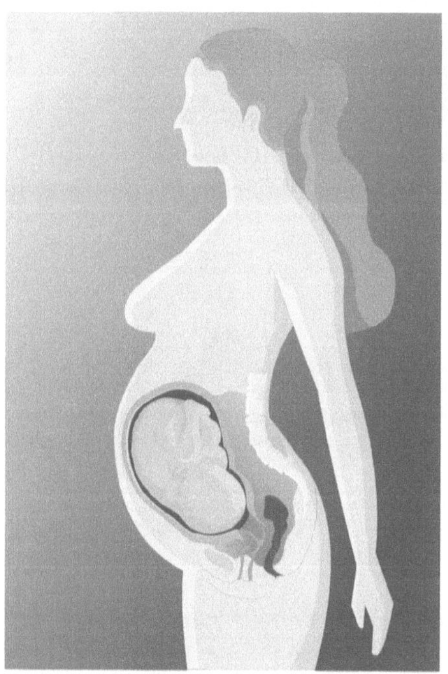

 Doctors in the *Medical Journals of Obstetrics and Gynecology* say that the most "secure age" for childbearing in women is twenty to thirty-five. (Anonymous, 2009)

 The childbearing age, which normally starts at fifteen, may start as early as age nine due to certain factors. It can continue until forty—or as late as fifty-nine. Birth of a child is blissful for parents and family. A mother experiences the miracle of gestation and growing the fetus in her womb, but the child's wellbeing depends on nutrition, attention, and loving atmosphere.

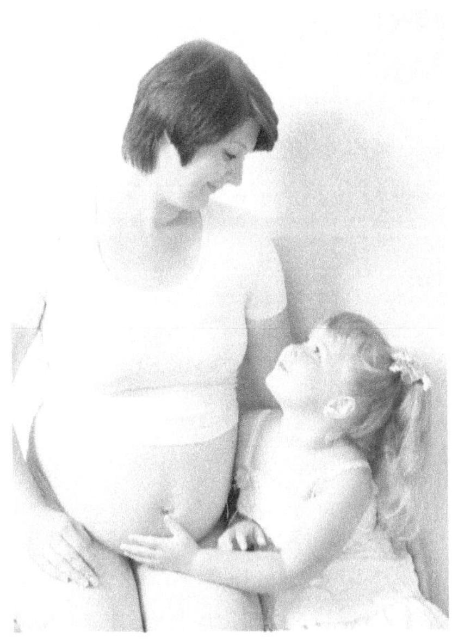

Menstrual Irregularities

They are manifested by not getting periods, getting periods too frequently, having unpredictable menstrual bleeding, or having painful periods. There are many conditions that can cause menstrual irregularities

Amenorrhea (No Periods)

No period by age sixteen or no period for at least three months. Causes are physical (excessive exercise), psychological (anorexia nervosa), or hormonal (polycystic ovary syndrome).

Case Presentation

Eighteen-year-old patient states she has never had a menstrual period.

Oligomenorrhea (Infrequent Periods)

An infrequent or irregular menstrual cycle with intervals of more than 35 days—but may vary. It is not a medical disease. Many women with polycystic ovary syndrome may experience oligomenorrhea.

Premature Ovarian Failure

This is a discontinuation in the normal functioning of the ovaries in women younger than forty. They may not have menstruation or may get periods irregularly. They may even become pregnant.

Dysmenorrhea (Painful Periods)

Dysmenorrhea including severe menstrual cramps can be caused by infection, endometriosis, or ovarian cysts, but it is usually not serious.

Painful periods can be treated with heating pads or warm baths. OTC medications can help relieve the pain. Birth control pills or shots from a health care provider are another option.

Case Presentation

Twenty-eight-year-old patient complained of severe pain with her menstrual periods, which had lasted for seven months. She had never had this type of pain previously.

Treatment of Irregular Periods

Lifestyle changes, Alternative Medicines, Drugs and Surgery (Love 2003, 34 Menopause Symptoms).

Lifestyle Changes

The first level of treatment available for women who are suffering from irregular periods is lifestyle changes. It is very important to take responsibility for changing your lifestyle and taking control. It involves weight reduction, stress management, introduction of some sort of physical activity, decreasing intake of refined carbohydrates, augmenting nutrition with multivitamins, calcium, magnesium supplements, fish oil, medications, oral contraceptives, and estrogen-replacement therapy.

These women should also reduce alcohol and coffee intake, consider meditation, yoga, any method of stress relief, consume lots of water, and increase consumption of fruits and vegetables.

Alternative Medicine

While changes in lifestyle will bring positive results, it is important to address the real cause of irregular periods—hormonal imbalance. In this approach, alternative medicines can be combined with lifestyle changes in order to get more satisfactory results. This level of approach can involve several different therapies, such as herbal supplements, biofeedback, massage, homeopathy, or hypnosis.

Drugs and Surgery

Since there is high risk and high costs with this type of intervention, only women with serious health conditions, such as uterine fibroids, needing a hysterectomy, or full removal of the female reproductive organs need this kind of treatment.

Hormone Replace Therapy (HRT)

This drug therapy is popular in West for treating irregular periods. This is a quick way to combat hormonal imbalance. There is the risk of serious side effects, including certain types of cancer, such as ovarian and breast cancer (Love 2003, 34 Menopause Symptoms).

There are few alternative approaches to treating irregular periods. Herbal supplements are used to stimulate or support hormone production. In acupuncture, fine needles are used in energy pathways to cure ailments. In homeopathy, small amounts of natural substances are used to stimulate the body's own defenses and healing processes. From these approaches, women find that herbs are the easiest alternative treatment to follow (Love 2003, 34 Menopause Symptoms).

In the case of herbal supplements, herbs that contain no estrogen are used to stimulate a woman's hormone production by nourishing the pituitary and endocrine glands, causing them to efficiently produce more natural hormones. This process results in balancing estrogen, progesterone, and testosterone.

Macafem is the best example for treating irregular periods; the body naturally creates its own hormones. Macafem stimulates a woman's natural hormone production by inducing the optimal functioning of the pituitary and endocrine glands (Love 2003, 34 Menopause Symptoms).

Partridge Berry

Available in tincture, powder, and capsule forms, it can be used combined with cramp bark for dysmenorrheal (twenty to thirty drops of each, three times a day, for dysmenorrheal periods).

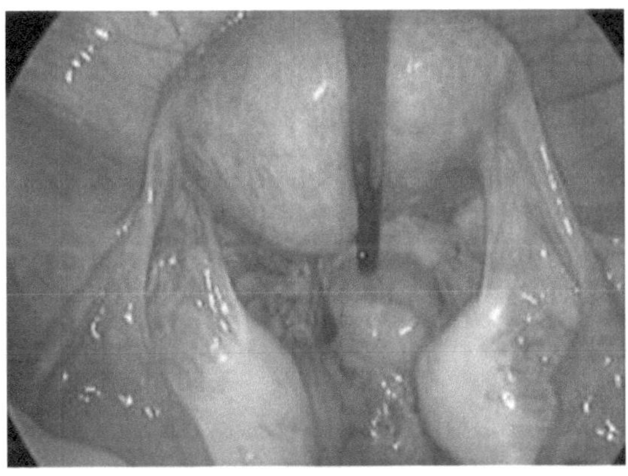

Endometriosis

Endometriosis is commonly seen in women between the ages of twenty -five and thirty-five. Endometriosis is among the health problems that affect these women. Women with this type of condition suffer from severe menstrual cramps, irregular periods, excessive bleeding, and painful sexual intercourse.

In this condition, the uterine tissue migrates outside the uterus to the ovaries, fallopian tubes, cervix, bowel, bladder, and sometimes the lungs. This condition occasionally causes infertility in women. The pathological condition of endometriosis is stimulated by estrogen.

Endometriosis is accidently diagnosed by the doctor when the patient is initially visiting for some other cause; an example is infertility (White and Foster 2000, 231).

Case Presentation

A twenty-four-year-old patient (eight months postpartum) reports excruciating back pain in the sacral area as well as left lower quadrant pain. The pain worsens about one and a half weeks prior to her periods and continues during her menses. She gets mild relief with NSAIDS. Since the delivery, she has been taking oral contraceptive pills; the pills regulate her cycles, but they do not seem to alleviate the pain.

Women with endometriosis may have very heavy periods. It can cause pain before and during the first few days of the menstrual period. About 50 percent of women with endometriosis are infertile, making it one of the top three causes of female infertility (Love 2003, 34 Menopause Symptoms).

Drug Treatment

Oral contraceptives and synthetic progesterone are widely used to treat this entity. Side effects are dizziness, headaches, weight gain, and insomnia.

Prevention

Cut down on fats and dairy products, especially hydrogenated oils and margarines. Phytoestrogens found in soy-based foods and beans can help prevent natural or synthetic estrogens. Phytoestrogen will decrease overall estrogen levels in your body. Reduce chocolate, caffeine, and sodas because they may worsen symptoms of endometriosis.

Natural Remedies

Natural Progesterone is available in creams, capsules, and vaginal or rectal suppositories, which stops the ovulation and controls estrogen level (White and Foster 2000, 230). Natural progesterone is synthesized from the wild yam.

Antioxidants, such as vitamin E (400-800IU), vitamin C (2,000 mg), grape seed or green tea extract (200-400 mg), and magnesium (500 mg) should be taken daily.

Chamomile, a calming herb with antispasmodic properties, helps reduce cramps. Typical dose is 300-400 mg capsules per day or ten to forty drops of tincture three times per day. Watch for allergic reactions, which some people may experience.

Cramp bark, a uterine sedative and tonic, which is used to relax uterine muscle, treats menstrual cramps, and may prevent miscarriage. Do not use this herb if you have kidney stones.

Wild yams are used in Ayurvedic and Chinese medicine. The root has an anti- inflammatory property. Typical dosage is 400 mg capsules per day or twenty to forty drops of tincture up to five times per day.

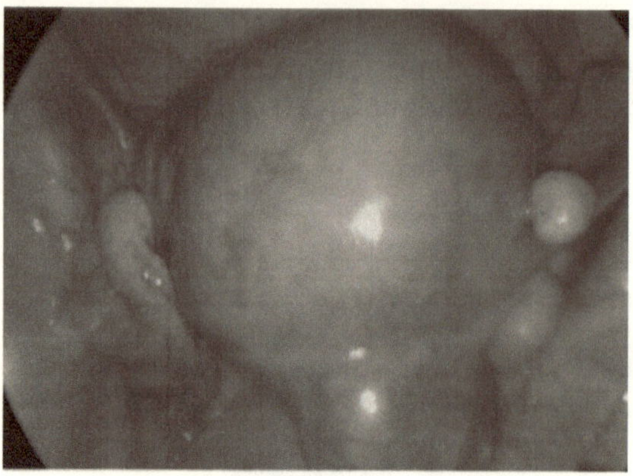

Fibroids

Fibroid is a common female health issue. It is a benign solid mass that occurs in the uterus during the ages of thirty-five to forty-five. It occurs in women who have never been pregnant. This is not a cancerous tumor. It is a slow-growing mass that causes heavy menstrual bleeding, bleeding between periods, a sense of heaviness in the pelvis, frequent urination, sudden severe cramps, and infertility (White and Foster 2000, 248).

Fibroids may prevent pregnancy and may cause miscarriages. High estrogen levels in the body, such as during pregnancy or with oral contraceptive use, cause fibroid growth. Other factors that cause fibroid growths are obesity, alcohol use, a high fat diet, and vitamin B deficiency.

Case Presentation

A forty-two-year-old woman complained of heavy bleeding during menses, which lasted for two years.

A thirty-five-year-old woman came to my office for a routine physical exam. She had no other complaints besides weight gain over the past year. Her weight was 243 pounds, her vital signs were normal, and she appeared in no acute distress. The health questionnaire she filled out did not reveal any complaints other than was not being able to conceive inspite of multiple attemps.

Drug Treatment

Gonadotropin-releasing hormone is available in natural form. Side effects include hot flashes, increased risk of heart disease.

Natural Therapy

Easy Pain Relief

Hot sitz baths will increase circulation and relax tight muscles. Add essential oils such as rosemary, lavender or juniper to the sitz bath (White and Foster 2000, 248).

The skin can absorb warm castor oil's active ingredients and lectins which stimulate the immune response to help shrink fibroids.

Milk thistle helps shrink fibroids by stimulating liver repair. Using the tincture form, take twenty -five drops three times a day for three to four months. Other herbs, such as dandelion and yellow dock help metabolize estrogen in the body.

Burdock has shown antitumor capabilities in animal studies. Take twenty -five drops three times a day and continue for three to four months (White and Foster 2000, 250).

Uterine fibroids

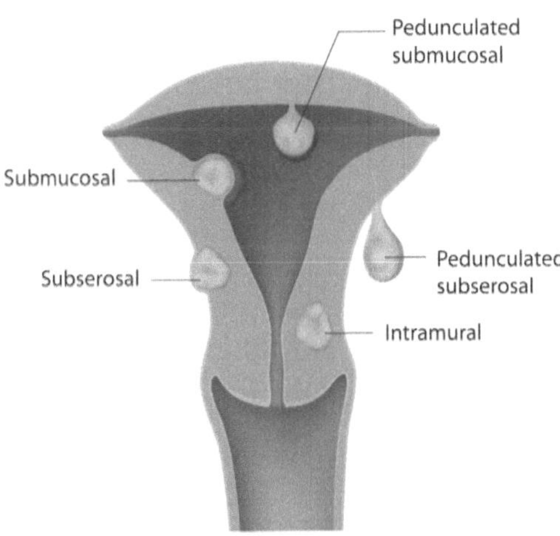

Shepherd's purse

Miscarriage

Miscarriages are the most common types of pregnancy losses. Most miscarriages occur in the first thirteen weeks of pregnancy. Many are due to inadequate diet, stress, and trauma. During the first trimester, the most common cause of miscarriage is chromosomal abnormality, meaning that something is not correct with the baby's chromosomes.

Other causes of miscarriages include hormonal problems, infections, maternal health problems, lifestyle (toxin exposure, smoking, use of drugs, alcohol, malnutrition, diet, and environmental problems), maternal age (women under age thirty-five have 15 percent chance of miscarriage, ages thirty-five to forty-five have 35 percent, women over forty-five have 50 percent chance), and maternal trauma (American Pregnancy Association, "Promoting Pregnancy Wellness," (http://www.nlm.nih.gov/ medlineplus/pregnancyloss.html).

Warning Signs

Warning signs include back pain, weight loss, white-pink mucus discharge from the vagina, true contractions, brown or bright red bleeding, and tissue with clot- like material passing through the vagina, and sudden decreases in signs of pregnancy.

Treatment

Since the cause is the chromosomal abnormality, nothing much can be done. Psychological support is very important once a miscarriage occurred.

Prevention

The Ayurveda principle advises always keeping your chest clean. *Kapha dosha* is involved.

Always try to keep a healthy lifestyle. Do not smoke or be around smokers. Do not drink and avoid environmental hazards, such as x-rays, radiation, and infections. Check twice before you take over-the-counter medication.

Herbal Treatment

Herbs like raspberry, squaw vine, and nettles can be taken during pregnancy; these herbs help tone the tissue and facilitate the birth itself. Herbs which act as *emmenagogues* and stimulate the uterus may trigger miscarriages. Barberry, goldenseal, pennyroyal, pokeroot, sage, wormwood, and rue should be avoided during pregnancy.

Rue

Certain herbs can provide extra strength and vitality and can help prevent unnecessary miscarriage. Black cohosh (two parts), false unicorn root (two parts), and cramp bark (one part) should be used three times daily. If stress is involved, skullcap or valerian can be added. Blue cohosh can also be used for this purpose. Please refer to *Materia Medica* for the blue cohosh description.

Red raspberry leaf, nettles, alfalfa, and dandelion are safe and easy to take during pregnancy.

Cervical Dysplasia

Cervical dysplasia is common in teenagers because of promiscuous practices and most often affects sexually active women under age twenty.

This premalignant lesion is associated with human papillomavirus (HPV) infection, which can turn into a cancer.

Case Presentation

A thirty-two-year-old patient said, "My husband and I have been married for ten years, and we have been trying to get pregnant for the past three years. I found out that I had pre-cervical cancer and had laser surgery to get rid of the bad cells."

Cervical dysplasia is a condition that changes to cervical cancer if not diagnosed early. Herbs that support and repair the liver are important for fighting cancer. Cigarette smoking, waste products such as bacteria and other organisms, and environmental toxins such as pesticides are major offenders. A decrease in liver functions has been linked to an increase chance of developing cervical dysplasia and cancer (White and Foster 2000, 160).

Pap Smears

Any woman who is sexually active or is at least eighteen years old should have this screening test for detecting cancerous or precancerous changes of the cervix. Cervical cancer is most commonly caused by human papillomavirus.

Drug Treatment

There is no drug that treats cervical dysplasia or reverses its progression to cancer. The use of ibuprofen, an anti-inflammatory drug, is being studied. One new drug, difluoromenthylornithine (DFMO), shows some promise. This drug seems to prevent its conversion to cancerous cells.

Prevention

Human papillomavirus causes genital warts and herpes simplex type II that causes genital herpes are known to be the causal factors for cervical dysplasia. Women who drink heavily and take oral contraceptives are prone to reduce the folic acid, a part of the vitamin B complex, in their bodies. Excess estrogen also increases the chance of cervical dysplasia and cervical cancer (White and Foster 2000). Avoid cigarette smoking. All women should have a regular pap test, which prevents the cervical cancer by early detection of abnormal cells.

Other factors that are reported to reduce the cervical dysplasia are adding beta-carotene, and vitamin A, vitamin C, and folate supplements (Giller and Matthews 1994).

Vitamin Supplements

Certain supplements are needed for patients with cervical dysplasia who are deficient:

- vitamin A (5,000-10,000 IU)
- riboflavin (10 mg)
- vitamin C (1,000-2,000 mg)
- folic acid (400-600 micrograms)
- vitamin E (400-800 IU)

You can also take lycopene, a carotenoid found in fresh or cooked tomatoes (1-5 mg daily). Try to purchase organic tomatoes to avoid chemical toxins.

Herbal Remedies

Many studies have verified the effectiveness of milk thistle's main compound, silymarin, which deactivates harmful free radicals. Milk thistle can be used in capsule form, starting with a 140 mg dose of standardized preparation, three times a day, reducing the dose to 90 mg three times a day after six weeks, or twenty-five drops of tincture three times a day.

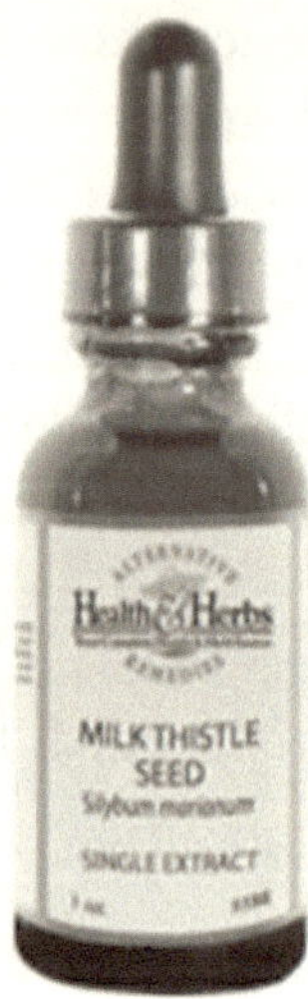

Red clover is a rich source of phytoestrogens; these plants are chemicals similar to human estrogen. The plant-based estrogen locks on to certain cells, preventing the real estrogens from overstimulating the body and stops converting them into cancerous cells (White and Foster 2000, 200).

Burdock, yellow dock, turmeric, ginger, echinacea, astragalus, reishi, and shiitake seem to affect cervical dysplasia, but no promising result has been noted.

Tea for Cervical Dysplasia

- 2 teaspoons *Vitex* berries
- 1 teaspoon burdock root
- 1 teaspoon red clover
- 1 teaspoon astragalus root
- stevia leaf (optional)
- 1/2 to 1 teaspoon peppermint, spearmint, or wintergreen (optional)
- 5 cups water

Boil herbs and water, simmer for five minutes, let steep for twenty minutes, and strain out herbs. Drink two or three cups per day. You can store this mixture in the refrigerator for up to three days.

Cervical Cancer

It is one of the most common cancers of the female reproductive system. Because there are no signs or symptoms in its early stages, it grows unnoticed. The symptoms of cervical cancer include irregular menstrual cycles, bleeding between periods, bleeding after menopause, bleeding after sexual intercourse, a heavy discharge from the vagina, and persistent pelvic pain.

The causal factor is the human papillomavirus, which has been linked to the development of cervical cancer. After initial infection with HPV, the virus remains in a dormant phase until, in progressive stages; it becomes dominant and causes cervical cancer.

Risk Factors

- having sex at early age
- many sexual partners
- unprotected sex
- infection with more than one STD
- compromised immune system
- smoking

Conventional Treatment

In a conventional approach, a combination of treatments, such as chemotherapy, radiation therapy, surgery, hormone therapy, and immunotherapy is considered. Medical providers can use one or a combination of the above treatments to treat the cancer.

In chemotherapy, toxic drugs are administered to destroy the cancer cells. A combination of two or more chemicals is given before or after surgery or in combination with radiation to improve survival and cure rates.

Radiation will cause the cancerous cells to shrink after introducing the high-energy x-rays to the target organ. It can be used before or after surgery to improve outcomes.

In surgery, a hysterectomy is done, during which the cervix and the uterus are removed.

In hormone therapy, an attempt is made to prevent the adherence of estrogen to tumor cells. Certain cancers are grown in the presence of estrogen. Tamoxifen is one drug that causes the blockage of the receptor site where normally estrogen is bind.

In immunotherapy, this treatment boosts the body's own defenses to destroy cancer cells.

Prevention

The best way to prevent these cancers is to make healthy lifestyle choices, such as limiting exposure to smoking. All women should adhere to safe sex practices to limit the possibility of contracting a sexually transmitted disease, such as HPV. The use of birth control pills and tubal ligation can decrease the likelihood of some female reproductive cancers. An HPV DNA test identifies the presence of the genetic material of HPV in cervical cells.

Diet and Supplements

Always try to consume organic foods, fruits and vegetables that are rich in antioxidants, whole grains, high-quality proteins, and fats that help the body function optimally.

Antioxidants such as vitamins C and E, high-quality medicinal mushrooms, and melatonin have been shown in several research studies to improve outcomes on some solid tumors. Antioxidants should be avoided while the patient is in radiation or chemotherapy treatment (*The Duke Encyclopedia of New Medicine* 2006, 414).

Alternative Therapy

Many supportive therapies are available, such as the ability to fight cancer with the body, mind, and spirit. Be prepared to face the physical struggle—and try to lift your emotional status and inner strength through positive relationships with family and friends. Keep a supportive spiritual foundation that will strengthen the immune functioning and well-being.

Breast Cysts

Case Presentation

A forty-nine-year-old patient presented a report that showed a presence of a breast lump. Examination done by her primary care provider confirmed a right breast upper outer quadrant three-centimeter, discrete, nontender, cystic mass.

It is a noncancerous, very painful lesion of the breast that affects women in their thirties and forties. This lesion is different from fibrocystic disease, which is chronic but benign. Cysts are filled with fluid due to hormonal changes. This fluid is transported by the lymphatic system. More cysts are formed due to the blockage in the lymphatic channels and in the circulation. Cysts are not dangerous by themselves, but health care providers should check them regularly.

Drug Treatment

Danazol shrinks breast lumps by decreasing the levels of follicle-stimulating hormones and luteinizing hormones, thereby reducing estrogen production.

Side effects are weight gain, acne, increased body and facial hair, voice changes, and menstrual disturbances (White and Foster and the staff of Herbs for Health).

Herbal Remedies

For dosing and detail, please refer to *Materia Medica* under Lady's Mantle.

Evening primrose oil contains essential fatty acids (natural antiprostaglandin) that help reduce breast lumps. Typical dosage is 1,500 mg in capsule form two times per day.

Two or more cups of 500 ml of violet leaves infusion, used daily, helps the soreness of the breast, monthly breast swelling, tenderness, fibrocystic disease, and mastitis (Weed 1989, 244).

Breast Cancer

Case Presentation

A forty-five-year-old black female with a complicated and advanced case of breast cancer and metastases to lungs and to the brain, status postmastectomies, was admitted to the hospital with a chief complaint of abdominal pain. The pain had gradually increased in intensity over one month. She described the pain as sharp and deep inside her abdomen; the pain was located centrally. She denied having diarrhea. The patient graded the pain as 9/10 and was accompanied by nausea and vomiting. She also complained about headaches.

She was receiving radiation therapy for metastasis to the lungs and brain. Mastectomies, hysterectomy, and two cesarean sections had been done in the past. Her medications were Zofran, Prilosec, Lisinopril, Glipizide, Dexamethasone, and morphine. She was one of the twelve children, out of which ten were females. She had a family history of maternal breast cancer. Six of the sisters had a breast cancer (Brown 2005).

According to www.breastcancer.org, about 15 percent of women who get breast cancer have had a family member diagnosed with it.

In other words, about 85 percent of breast cancers occur in women who have no family history of breast cancer. The cause in these cases is said to be a genetic mutation that happens as a result of the aging process. It has been estimated that one out of eight (12 percent) of women in this country will develop breast cancer in their lifetimes. It is one of the most common cancers in women—and the second-most common cancer death for women (Clark 2000). The cancerous cells spread to the other parts of the body and invade the liver and bones in advanced stages.

Breast pain is a prominent symptom in breast cancer. It is diagnosed by feeling a hard, nontender mass within the breast, which is not moveable.

Risk Factors

Not all women have the same risk of developing breast cancer (Clark 2000). The following factors increase risk:

- previous history of breast cancer
- family history of breast cancer (mother/sister)
- early onset of menstrual period
- late menopause
- childless or having a first child after age thirty
- high-fat diet
- obesity
- alcohol use (nine drinks per week)
- radiation exposure
- prolonged estrogen and progesterone use (controversial)
- urban lifestyle

Note: Breast-feeding mothers are at lower risk for breast cancer.

Conventional Treatment

Some of the conventional methods for the treatment of breast cancer are radiation therapy, chemotherapy, hormonal therapy, and surgery. All of them have side effects, but using these conventional lines of treatment with herbal remedies—especially with Ayurvedic medicines—gets faster results with minimal side effects.

Prevention

It is important to make healthy lifestyle choices. This includes maintaining a healthy weight, eating fruits, vegetables, lean meat, and foods high in fiber, regular physical activity, managing stress, and limiting exposure to smoking and pesticides. Cooking with olive oil is important because it contains healthy monounsaturated fats. Use safe sex practices to limit the exposure to sexually transmitted diseases, such as HPV.

Diet and Supplements

Focusing on organic foods, fruits and vegetables rich in antioxidants, whole grains, and high-quality proteins and fats helps the body fight with the disease. Consumption of items containing beta-carotene (vitamin A) and vitamin C, which are used as antioxidants, and a diet low in saturated fats are also very beneficial.

Alternative Medicine

Support the body, mind, and spirit's ability to fight cancer. Strengthening the role of family and friends in facing struggles and questions of mortality, finding peace, and supportive spiritual foundation can strengthen *ojas* (immune functioning) and well-being.

Herbal Treatment

High-quality medicinal mushrooms may help in some cancers (*The Duke Encyclopedia of New Medicine* 2006, 414).

Other herbal remedies that are beneficial and help reduce the severity of breast cancer are garlic, burdock root, alfalfa, and red clover. According to research done by the National Cancer Institute (Clark 2000), anticarcinogenic compounds are found in red clovers.

Bioflavonoid and ligans are other sources for lowering the risk of breast cancer.

Ayurveda

When dealing with a health condition in general—and with cancerous cell in particular—always keep three things in mind: *ojas* (immunity), *tejas* (intelligence), and *prana* (life force). According to Ayurveda, breast cancer is caused due to the imbalance in three *dosha* s (*vata, pitta,* and *kapha).* Therefore it is called *tridosha.* There is increase in accumulation of air (*vata*) manifested by anger and frustration and an increase in toxins and heat (*pitta*) in the body because of foods with pesticides, free radicals, use of aluminum, plastics, lots of fat, salt, spices, and animal products. The increase in *kapha,* manifested by an accumulation of water in the body, causes toxins that trigger the cancerous cells in those who have decreased immunity and are prone to a proliferation of abnormal cells.

Ayurvedic Treatment

The Ayurvedic method of treatment consists of *Panchkarma* (cleans *vata, pitta,* and *kapha). Yukti Vyapashraya* is used to get rid of dead cells in the patient's body, treating breast cancer. *Rasa Rasayan* treatments get rid of toxic elements from cancerous cells. *Rasa Rasayan* treatments are used to compensate for body elements that have become deficient, such as mercury, copper, iron, zinc, sulfur, and silver.

Ayurvedic therapy acts on *mansa* (tissue), *rasa* (taste), and *rakta* (blood), which are effective in breast cancer treatments. The Ayurvedic herbs include *Shankh-Vati, Ashwagandha, Shatavari, Tulsi, Bhringraj, Nimba, Triphala,* and *Kutki,* which are also effective for immune dysfunction.

Ovarian Cysts

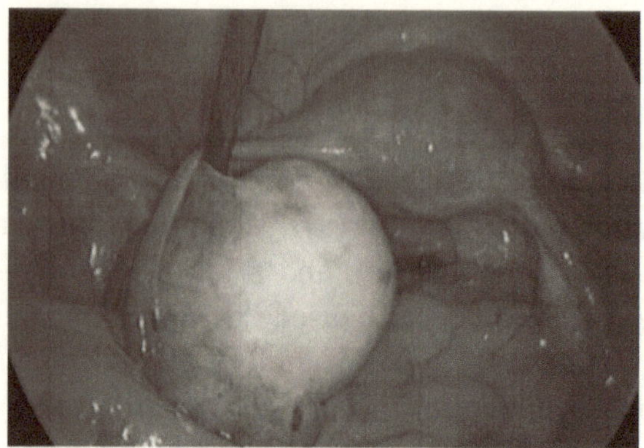

Case Presentations

A seventeen-year-old Asian girl came to my office and complained of a pain in the groin region on both sides. The girl had painful menstruation, which lasted about one week, and the pain lasted for at least six months. The bilateral ovarian cysts were confirmed by sonography with her primary care provider. She refused to go for any surgical procedure recommended by her physician. Eventually, she was referred to me for herbal advice and was managed with herbal treatment for four months. She totally became free of any symptoms and menstrual pain. After being free of pain after nine months, she resumed her regular menstrual periods.

A twenty-five-year-old patient complained of pelvic pressure symptoms. The physical report from her primary care provider showed a five-centimeter cystic adnexal mass on the right side of her pelvic area.

Ovarian cysts are mainly functional cysts common in women in childbearing age. The ovaries get filled with fluid and become quite large over time. Luckily, natural remedies can get rid of those cysts without surgery.

Initially, there are no signs or symptoms. They are discovered on routine physical exams.

The symptoms may include pelvic pain, especially after intercourse or strenuous exercise, usually on the side where the cyst is present. I have seen more than twenty cases, mostly unilateral pelvic pain in women under thirty-five, and a few cases with bilateral pelvic pain. The pain may be sharp and sudden or come and go throughout the cycle.

Conditions that Cause Ovarian Cysts

- hormonal imbalances
- irregular menstrual cycle
- hypothyroidism
- cigarette smoking
- early menstruation (eleven years or younger)
- increased upper body fat
- infertility treatment with gonadotropins
- use of Clomiphene
- use of Tamoxifen for breast cancer

Symptoms include nausea, vomiting, vaginal pain, pressure and/ or abdominal bloating, and pain or pressure in the abdomen when urinating or having a bowel movement.

Medical Treatment

Most commonly, birth control pills are used to prevent ovulation. The problem with this treatment is that it also causes long-term hormonal imbalance. Therefore, it does not solve the root cause. Surgery is performed if the cyst becomes too large, and the cyst twisting or adhering the other parts of the body is possible.

For functional cysts, there is a three-step approach (Rodriguez, 2012).

Step One

Excess estrogen and cysts should be cleared from the body; this treatment has no side effects. Reduce estrogen by reducing or eliminating your exposure to xenoestrogens and taking a supplement called DIM (a natural remedy for balancing estrogen).

Stop eating soy foods; eat only organic meat and dairy. Do not microwave foods in plastic; stop drinking water from plastic bottles. Avoid mineral oils and perabens in skin care products; use natural detergents.

DIM helps clear the body of these excess estrogens, aiding in hormonal balance and reducing the promotion of cyst growth.

Step Two

Increase the amount of progesterone in your body. When there is excess estrogen in the body, there is usually a progesterone deficiency. Therefore, by taking natural progesterone, you will reduce the ovarian cysts. Natural progesterone cream is available for this purpose; take it from days ten to twenty-six of the cycle. The cream will stop ovulation. No follicles occur, and no ovarian cysts can be created.

Step Three

There are two natural therapies that have been used to help the body break down the cysts, reduce their sizes, and help them to disappear.

Fertility enzyme therapy contains a specific enzyme that breaks down foreign tissues in the body. It eats away the cysts by reducing their size over time. By reducing the size of the ovarian cysts, you also get rid of excess estrogen. This can be done with the estrogen metabolizer DIM and progesterone.

Castor oil packs are ancient therapies that help cleanse and heal the body wherever they are placed. They clear the body of excess tissues and toxins. Castor oil packs stimulate the lymphatic and circulatory systems. The lymphatic system removes toxins and wastes from the area stimulated by the castor oil packs. Do not used castor oil packs when you are trying to conceive.

Use the castor oil packs for thirty minutes a day at least six days of the month at any time after the menstruation; apply for a total of three months.

Dosages of Natural Therapy

- DIM-one capsule three times a day
- Maca-two tablets a day
- Wobenzym-three tablets two times a day between meals

From days 10 -26 of your menstrual cycle, use progesterone cream (20-40 mg a day), starting with 20 mg and increasing the dose to 40 mg. Use this method of progesterone therapy for three months and then stop; this is used to suppress ovulation. If no ovulation occurs, no follicle occurs—and no ovarian cysts can be created. Do not use progesterone in this way when you are trying to become pregnant.

Supportive Herbs for Preventing Ovarian Cysts

After the symptoms have been covered with the above treatment, we attempt to help balance the body from preventing future cysts.

Lily flowers, nettle, red raspberry, and Queens Anne's Lady are used as contraceptives to correct the condition. Nutrients including vitamin B (B1, B6, B12), magnesium, and zinc, supported by diet, change in lifestyle, and psychological issues such as fear need to be addressed (Kress 2005).

We like to use the herbs to help nourish the endocrine system, promote hormonal balance, regulate ovulation, and promote proper circulation in the reproductive organs. These herbs also detoxify the liver for improved hormonal balance. The herbs below have been found to be supportive for proper menstrual cycle, reduction in ovarian pain, and increases in circulation to the reproductive organs and support the liver function.

- Maca root (*Lepidus meyenii*): Helps the body produce progesterone. Daily dose is 2,000-3,000 mg.
- Black cohosh root (*Actaea racemosa*): Promotes regulation of the entire menstrual cycle and is excellent for relieving ovarian pain.
- Dong quai root (*Angelica sinensis*): Supports healthy circulation to the reproductive organs, also aids in reducing pain.

- Milk thistle seed (*Silybum marianum*): Helps the liver get rid of toxins, including excess hormones.
- *Tribulus* (*Tribulus terrestris*): A nourishing herb for the female reproductive system, especially concerning the ovaries.
- *Vitex*, chaste tree berry (*Vitex agnus-castus*): Aids in regulating hormonal balance, promotes ovulation, and improves timing of the menstrual cycle.
- Wild yam root (*Dioscorea villosa*): Promotes a healthy menstrual cycle and reduces ovarian pain.
- Yarrow (*Achillea millefolium*): Relieves pelvic congestion and improves the timing of the entire menstrual cycle.

www.naturagenics.com 2012

Client Testimonies

R.S: After being on birth control for three months solely to treat my ovarian cysts, I was told by my primary care provider, that my next option was surgery. I knew there was another method of treatment out there, and I decided to seek herbal treatment. I took female balance herbal pills (wild yam, dong quai, and *Vitex*) for several months for my ovarian cysts and dysmenorrhea. Slowly but surely, my cysts and the crucial pain that came along with it disappeared.

In my seven years of herbal practice, I have seen and consulted many cases of ovarian cysts. Either they are related to genital infection from taking birth control pills or secondary to miscarriage. From the cases that have been referred to me, the real cause behind this pathology is still obscure. As far as their management is concerned, they have given a very good response and have become free of pain with herbal remedies and counseling for choosing healthy lifestyles.

Summary:

We have learned the main cause of functional cysts is hormonal imbalance, which disrupts the natural menstrual cycle and may lead to the formation of ovarian cysts. It is vital to bring the body back to a state of balance to prevent the formation of ovarian cysts.

- Reduce excess estrogen in the body by avoiding exposure to xenoestrogens and naturally promoting healthy estrogen metabolism through the use of DIM.
- Increase progesterone levels and support hormonal balance overall. Consider the benefits of natural progesterone supplementation and herbs for hormonal balance and reduction in cyst formation.
- Dissolve and reduce ovarian cysts through Fertility Enzyme Therapy (FET) and castor oil packs.

Rodriguez, 2012

Ovarian Cancer

Case Presentations (Pelvic Mass in Postmenopausal Women)

A sixty-year-old patient came to me for an annual follow-up. She was nine years postmenopausal (PM). She was diagnosed with a five centimeter solid mobile right adnexal mass.

A fifty-four-year-old patient came to my office complaining of abdominal enlargement and bloating. Her primary care physician found shifting dullness in her abdomen, and a pelvic exam revealed diffused masses.

- 25 percent of genital cancers are ovarian, but they result in 50 percent of deaths
- Ovarian cancer is the leading cause of death attributable to GYN cancers in the United States.
- 12 out of 1000 women will develop the disease; only 2-3 of the twelve women will be cured.
- Incidence rises in the fifth decade and continues until the eighth decade. The postmenopausal patient is at greatest risk.
- Ovarian cancer is silent in its early development; in 70 percent of the cases, the disease has spread beyond the pelvis by the time the diagnosis is made.

www.obgyn.uab.edu/medicalstudents

Patients must have annual pelvic exams during the postmenopausal years in order to detect ovarian enlargement.

Symptoms of Ovarian Cancer

Abdominal pressure or bloating, the frequent need to urinate, changes in bowel movements, pelvic and lower back pain, painful intercourse, and fatigue are common symptoms. Ovarian cancer often spreads to surrounding tissues and organs before it is noticed.

Who is at Risk?

Women over the age of forty, women of Caucasian origin with extended years of menstruation, women who have never been pregnant, women with irregular menstruation, obese women, diabetics, women who have received estrogen alone as hormonal replacement therapy, and women with family history of ovarian cancer.

Medical Treatment

Chemotherapy and whole abdominal radiation is recommended, and a 20 percent five-year survival rate is noted.

Prevention

It is important to make healthy lifestyle choices. These include maintaining a healthy weight, eating fruits, vegetables, lean meat, and foods high in fiber. Cooking with olive oil is important because it contains healthy monounsaturated fats. Regular physical activity, managing stress, and limiting exposure to smoking and pesticides is recommended. Use safe sex practice to limit the exposure to sexually transmitted diseases, such as HPV.

Diet and Supplements

Focusing on organic foods, fruits and vegetables rich in antioxidants, whole grains, and high-quality proteins and fats helps the body to fight the disease.

Alternative Medicine

Supporting the body, mind, and spirit's ability to fight cancer, strengthening relationships with family and friends in facing struggles and questions of mortality, and finding peace and a supportive spiritual foundation can strengthen immune functioning and well-being.

Herbal Treatment

High-quality medicinal mushrooms may help in some cancers (*The Duke Encyclopedia of New Medicine* 2006, 414).

Polycystic Ovary Syndrome (PCOS)

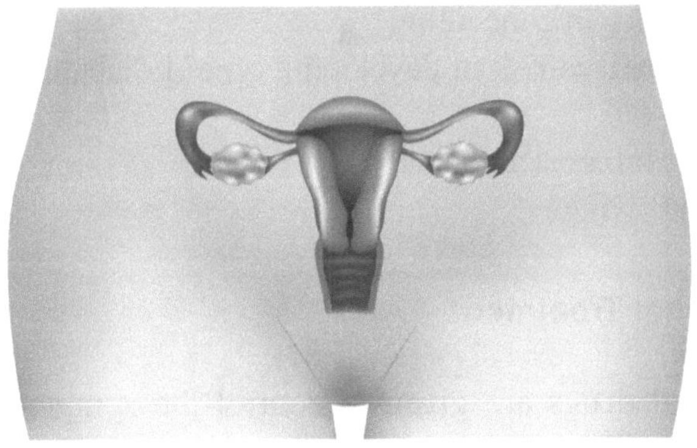

Introduction

By definition, Polycystic Ovary Syndrome (PCOS) is the malformation of the cysts inside the ovaries. The symptoms and ailments in a woman's life depend on the absence of menstruation in adolescence, infertility, and unexplained weight gain in post adolescence.

Polycystic ovary syndrome (PCOS) also called polycystic ovary disease (PCOD). PCOS refers to hormonal imbalances that involve many organs in the body and cause irregular menstrual periods, weight gain, increased body hair, breast shrinkage, acne, infertility, diabetes, high blood pressure, and cardiovascular disease. It is also reported that some women develop darkened skin around the neck, armpits, inner thighs, vagina, and beneath the breasts.

Because of the hormonal imbalance, the body produces an excessive amount of androgens and low levels of follicle stimulating hormone (FSH) that play a large part in the production of male type symptoms, such as acne and body hair. Ovulation does not take place regularly, eggs mature but are not released and cysts form inside the ovaries. This causes a large abdomen, pain in the abdomen, and change in bowel movements.

Common Symptoms

- weight gain
- excess body and facial hair
- difficulty in conceiving
- an increased risk of developing type II diabetes and/or heart disease
- skin problems and acne
- mood swings

Conventional Treatment

Hormonal drugs are given to control the symptoms, such as contraceptive pills containing estrogen and progesterone or androgen supplements. Short-term medications are given to induce ovulation in order to conceive and become pregnant. Clomiphene citrate is given to increase the likelihood of ovulation. Anti-androgens- spironolactone, a diuretic (this medicine used as a diuretic), is prescribed to block the effects of androgen, which will decrease unwanted hair.

Prevention

In order to keep an ideal weight, it is important to combine dietary supplements with a regular exercise regime. Whatever system you choose, be sure that you can commit yourself to a regular routine. Four thirty-five-minute sessions a week will be much more effective in helping you reach your target weight than one two-hour session of exerted activity a week.

Loss of excess weight should be a top priority because being significantly overweight increases the levels of androgens (male hormones that aggravate symptoms such as hair growth). It also increases the risk of complications such as type II diabetes and heart disease.

Complementary Treatment

Most naturalists or herbal practitioners will focus on restoring a healthy balance in the whole system. Being significantly overweight or having uncontrolled diabetes is known to aggravate PCOS. Certain measures can be used alongside conventional treatment.

Nutritional Management

Stay away from refined sugars to keep your blood sugar level in control. Sweetened drinks, sauces, and condiments contain large amount of hidden sugar. In order to keep your blood sugar level in balance, it is important to eat fruit, fresh raw chunks of vegetables, or rice cakes made from organic brown rice.

Avoid crash diets—especially low-calorie or low-fat diets. These do little to improve your nutritional status. Instead, eat regular small amounts of unrefined carbohydrates, such as whole grains, fish, fresh fruits, crisp vegetables, and moderate amounts of dairy.

Herbal Remedies

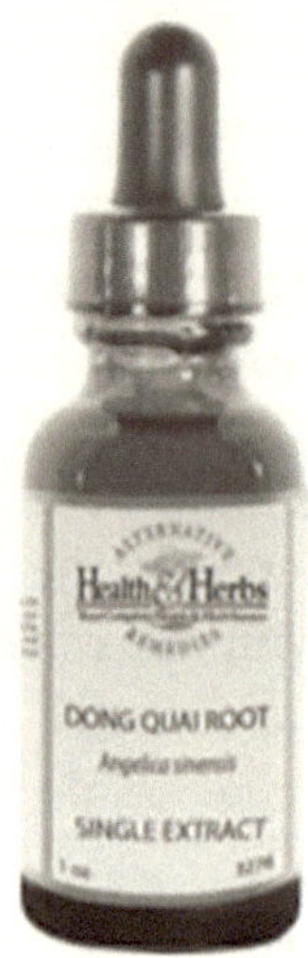

Dong Quai

Natural and holistic treatments have also proven to promote female reproductive health and support hormonal balance. A very good example of an herb used for centuries is dong quai; it's used as a nourishing blood tonic and helps maintain normal levels of estrogen and progesterone throughout the menstrual cycle.

One thing to keep in mind is to always consult a doctor before starting the herbs, especially if the client is using prescription medication, such as birth control pills or infertility medication, or has a preexisting condition, such as diabetes or high blood pressure. These herbs will not harm a person, but there are some that can increase or decrease the efficacy of certain medications. Some affect blood sugar or elevate blood pressure.

Vitex (Chasteberry)

This herb has been used for centuries for hormonal imbalance in women. It is one of the important herbs for PCOS because it helps stimulate the function of the pituitary gland, which controls the release of luteinizing hormone (LH). This herb indirectly raises progesterone levels in the body. *Vitex* is not recommended for those who are currently undergoing infertility treatments; it will increase the effects of infertility medications if taken with *Vitex*. Also avoid taking this herb while taking birth control pills.

Liver-Cleansing Herbs

Dandelion and milk thistle will remove the buildup of any toxins and excess hormones that are accumulated in the liver as well as healing damaged cells.

Saw Palmetto

Many women with PCOS have elevated levels of testosterone that are converted into Diihydrotestosterone (DHT). DHT is what stimulates the male pattern hair growth, and male pattern hair loss. Saw palmetto will help break down DHT. It has also been shown to have anti-estrogenic effects. In addition to reducing the overall levels of testosterone, DHT, and estrogen in the body, saw palmetto increases libido and helps with weight loss.

Infertility

Case Presentations

- A Primary Infertility: A thirty-year-old patient had been married for five years. In spite of not using any contraception, she was unable to get pregnant. She desperately wanted a child and came to my office for help.
- Secondary Infertility: A thirty-year-old patient had a successful term pregnancy three years earlier. In spite of not using any contraception, she had been unable to get pregnant again. She came to my office for further advice.

By definition, infertility is the biological inability of a woman in childbearing age to conceive. Different sources have different statistical figures in occurrences of infertility, but generally speaking, 40 percent issues involved with infertility are due to the woman, 40 percent are due to the man, and another 20 percent result from complications with both partners.

Infertility can be caused by urban lifestyle such as consumption of alcohol, smoking, high -fat diets, obesity, environmental toxins, infections of the fallopian tubes, and sexually transmitted diseases. Other conditions for infertility include age, ovulation disorders, elevated prolactin levels, prolapsed and fibroids of the uterus, cervical lesions, stress, and emotional problems.

The menstrual cycle is regulated by endocrine functions; because of endocrine abnormalities of the pituitary, thyroid or adrenal glands, ovulation does not take place.

Dr. Steinkampf said, "There could be something wrong with the ovary, the egg, the tubes, the uterus, or the sperm."

There could be a problem with ovulation, anatomical defects in the female genital tract, or abnormal spermatogenesis.

Medical Treatments

Fertility Medication

The most common treatment for female infertility is regulating or inducing ovulation. Clomiphene citrate helps the pituitary gland release FSH and LH to induce ovulation. Human menopausal gonadotropin (hMG) contains FSH and LH. This drug is used to stimulate the ovaries to produce more mature eggs (*The Duke Encyclopedia of New Medicine* 2006, 396).

Surgery can remove any blockage in the fallopian tubes. Assisted reproductive technology (ART) is called artificial insemination, in which the sperm is injected into the cervix or uterus. During in vitro fertilization, a fertilized egg is implanted within the uterus.

Prevention

Maintaining a healthy lifestyle—keeping healthy weight, a moderate exercise routine, limiting the use of medications and alcohol, avoiding smoking and illicit drugs, and practicing safer sex—will protect against complications from sexually transmitted diseases.

Vitamins and Supplements

Vitamins C and E have been shown to increase fertility in women. The amino acid L -arginine, when taken during in vitro fertilization, may increase the success rate of conception. Iron supplements may improve the chances of conceiving in those who are deficient.

Alternative Therapies

Acupuncture

Treatment involves the placement of fine needles in specific sensory locations on the body. Studies show that infertile women had increased success with conceiving after this treatment (*The Duke Encyclopedia of New Medicine* 2006, 396).

Herbal Treatment

Vitex, also known as chaste tree, can help correct deficiencies in the menstrual cycle and correct elevated levels of prolactin.

Chaste tree, wild yam, kelp, false unicorn root, and Shatavari are great herbs for fertility (Clark 2000). These herbs balance hormones in the body.

Chickweed is a powerful nourisher to the glandular and lymphatic system if the cause of infertility is an ovarian cyst, ovarian cancer, or another reproductive cyst. A fresh tincture of chickweed (forty drops, twice daily) is recommended until the desired effect is noted.

Seaweed is another good ally; it is used in infertility cases in same dose as chickweed.

Postpartum Depression

(1998-2012 Mayo Foundation for Medical Education and Research)

Observed Cases

- In May 1995, a forty-three-year-old female patient was admitted to the psychiatry ward at Temple University in Philadelphia. She suffered from delusional thinking, fifteen days after giving birth to her first child. She thought that her child was a demon from a different planet who wanted to kill her.
- Angela returned to my office six weeks after the delivery of her child. She stated that she had been feeling very tired and often found herself crying. She felt that caring for her baby was "overwhelming." Her husband was supportive but worked long hours and was not home much. Angela felt isolated at home but did not feel ready to return to work and

had requested an extended leave of absence. She and her husband had not had sexual intercourse since the birth of their child.

Introduction

Postpartum depression is still underestimated and neglected in medical care facilities. Powerful emotions can happen after the birth of a baby that range from excitement and joy to fear and anxiety. It can also result in depression and psychosis. Psychotic manifestations and delusional thinking are not uncommon.

Many new moms experience the "baby blues," which commonly include mood swings and crying spells. But some new moms experience a more severe, long-lasting form of depression known as postpartum depression.

Drug Treatment

SSRI such as Prozac, Paxil, and others are given in these cases. Side effects are anxiety, agitation, insomnia, and sexual dysfunction. An estrogen patch can be used to treat postpartum depression, but it decreases milk production.

Prevention

A new mother needs a very supportive and social environment. Adequate nutrition and vitamin intake during the prenatal period are very important.

Herbal Remedies

St. John's wort works for women suffering from mild to moderate depression.

Consult with your primary care provider if you are using prescription medications. 5-HTP is one of the raw materials that your body needs to make serotonin, which is necessary for alleviating depression. Omega-3 fatty acids and saffron are also used for depression. Flower essences are used for depression and anxiety and have given a very satisfactory response (Mayo Clinic 2010, 166).

Valerian, Hops, and Chamomile

These herbs are very effective in cases when a patient is suffering from fatigue, excessive anxiety over her child's health, difficulty concentrating or making decisions, or a loss of interest in daily activities.

Flower Essences

- unlocking the seals of the seven main chakras
- emotional and physical relief

Morning Sickness

Introduction

Most morning sickness commonly occurs in first few months of pregnancy. It occurs in the mornings when the stomach is empty due to changes in hormonal levels combined with low blood sugar and low blood pressure. It is best to avoid any medication during this period. Many women experience some degree of nausea during early pregnancy. In many cases, the symptoms will disappear by the end of the first trimester. Severe nausea and vomiting should not be ignored.

Drug Treatment

Several medicines are available to treat morning sickness during pregnancy. Emetrol is available as a nonprescription drug. Reflux medications such as Zantac or Pepcid are also used to control nausea and vomiting. Benedictine is a combination of Doxylamine—an antihistamine—and vitamin B6. It was widely used in the United States, but reports that it was harmful during pregnancy forced the manufacturer to pull it from the market.

Prevention

- Try to avoid any greasy foods.
- Have frequent protein snacks.
- Keep your stomach full.
- Apply pressure to the acupressure points on the wrist. Measure three-finger widths up the inside of the wrist from the crease and press for several minutes.

Herbal Remedies

There are some safe and specific remedies used for this condition:

Meadowsweet 2 parts
Black horse bound (1 part)
Chamomile (1 part)
Vitamin B6, (50 mg, take daily for seven days)
Ginger capsule (250 mg, three times daily)

Labor

Introduction

Labor is a physiologic process during which the products of conception (i.e., the membranes, umbilical cord, and placenta) are expelled outside the uterus. The onset of labor is defined as regular, painful uterine contractions resulting in progressive cervical effacement and dilatation of the cervix.

Medical Treatment

There are several treatment options available to induce labor, including epidural blocks, spinal blocks, narcotics, local anesthetic injections, pudendal blocks, tranquilizers, and nitrous oxide. Many types of medications can ease the pain during labor and delivery.

Each type of treatment has its pros and cons. They can cause decrease in sensation on one side of the body, decreases in blood pressure in the mother or baby, and respiratory problems.

Herbal Remedies

Angelica root is a uterine tonic and hormonal regulator that promotes estrogenic activity. It has been used in China for centuries for women's health; it helps expel a retained placenta. Use it as a tincture (ten to fifteen drops, under the tongue) to help expel the retained placenta.

Black cohosh is used for labor induction. Black cohosh can start contractions and should be used with caution during pregnancy since it may induce abortion.

Blue Cohosh is similar to black cohosh.

Drinking castor oil with juice is commonly called a "midwife cocktail" or castor oil induction. Castor oil induction is known to induce contractions after it stimulates the gastrointestinal system and causes diarrhea. When the bowel is stimulated in this way, the uterus is also stimulated, causing contractions. A study done in *Alternative Therapies* in January 2000 reported that 57.7 percent of the mothers given castor oil began labor in the specified time period, and only 4.3 percent of mothers not given castor oil went into labor. There are many recipes for castor oil drinks, but the one that has a pleasant taste is three teaspoons of castor oil mixed with a glass of orange juice and one teaspoon of baking soda.

Comfrey leaf is used externally to help prevent infection and promote healing. It can be helpful after the baby is born for perineal tears. It can reduce pain and prevent infection. It can be added to a sitz bath or used in a compress.

Add four ounces of dried comfrey leaf to half a gallon of boiling water and steep for eight hours. Use the liquid for sitz baths (fifteen minutes, twice a day).

Combine 1/4 cup comfrey, 1/4 cup uva ursi, and 1/4 cup shepherd's purse with water. Bring the mixture to a boil, reduce heat, and simmer for twenty minutes. The liquid can be used for compresses or as a wash for the perineal area.

Evening primrose oil is extracted from the seeds and placed in a capsule. It helps soften the cervix by providing prostaglandins. This can help ripen the cervix to help encourage labor or speed labor.

Evening primrose oil can be taken internally near the end of pregnancy. It can also be massaged directly onto the cervix if ripening is desired. Evening primrose oil should not be taken until after thirty-six weeks of pregnancy.

A tea made from the leaf of the red raspberry can help tone the uterus throughout pregnancy. It also helps increase fertility and reduces the risk of miscarriage. Some women find it eases morning sickness, and it is believed to help make labor more efficient and less painful.

During labor, it helps stimulate stronger contractions. After the baby is born, it helps milk production and assists the uterus in returning to prepregnancy size.

If the strength of uterus during partition is weak and labor is prolonged, goldenseal works as an oxytocic herb. It should not be used during pregnancy. Its use during difficult labor strengthens the mother's body to facilitate the labor.

Milk Production

Introduction

Every woman who is prepared to become pregnant and give birth should have essential knowledge about breastfeeding and natural nutritional needs of the infant.

During pregnancy and the first few months of postpartum, a woman's milk supply is controlled by hormones. Essentially, as long as the proper hormones are in place, the mother will start making colostrum about halfway through pregnancy—and her milk will increase in volume around thirty to forty hours after birth.

During the later part of pregnancy, the breasts are making colostrum, but high levels of progesterone inhibit milk secretion and keep the volume down. At birth, the delivery of the placenta results in a sudden drop in progesterone/estrogen/HPL levels. This abrupt withdrawal of progesterone in the presence of high prolactin levels increases milk production (Breastfeeding Basics, Archives 2011).

Research shows that milk volume is typically greater in the morning hours (a good time to pump if you need to store milk) and falls gradually as the day progresses.

As far as the milk supply is concerned, milk is being produced at all times; the speed of production depends upon how empty the breast is. Milk collects in the breasts between feedings, and the amount of milk stored in the breast between feedings is greater when more time has passed since the last feeding. The more milk in the breast, the slower the speed of milk production

To speed milk synthesis and increase daily milk production, the key is to remove more milk from the breast, quickly and frequently, so that less milk accumulates in the breast between feedings.

Cultural, social, financial issues can affect milk production during pregnancy. It is best for the child to be on breast milk for as long as it is feasible. According to WHO, breastfeeding helps the mother and the infant for optimal survival and longevity and society at large. In other words moms that breastfeed their children help them grow to be healthier than those kids who aren't breast fed. Breastfeeding is recommended at least six months to provide the best nutritional source and avoid malnutrition in the infant. Breastfeeding is the normal way of providing infants with the nutrients they need for healthy growth and development. With accurate information, virtually all mothers can breastfeed— provided they have the support of their family, the health care system, and society at large.

Colostrum, the yellowish, sticky breast milk produced at the end of pregnancy, is recommended by WHO as the perfect food for newborns, and feeding should be initiated within the first hour after birth.

Breastfeeding is one of the most effective ways to ensure child health and survival. Optimal breastfeeding and complementary feeding helps prevent malnutrition and can save about a million child lives. When breast milk is no longer enough to meet the nutritional needs of the infant, complementary foods should be added to the diet of the child.

Herbal Supplements

Herbs that can help produce milk are aniseed, blessed thistle, fennel seeds, fenugreek seeds, goat's rue, and vervain.

Ayurvedic herbs, such as shatavari, are given to increase milk production 2 teaspoons twice a day for four to six weeks.

Increasing Milk Production

Massage Oil

- 15 drops of fennel, geranium, or clary sage
- 2 tablespoons carrier oil
- Massage breast in a circular motion.

Use this formula in tea or tincture form.

- blessed thistle (1part)
- raspberry leaf (1 part)
- borage (1/2 part)
- nettles (1/2 part)
- fennel seed (1/4 part)

St. John's wort can be used. If the milk flow is to be stopped, the most effective herb is red sage. If it is not available, goldenseal can be used until desired results are obtained.

Summary

You learned about the following herbs in this section: Blessed thistle, Raspberry leaf, St. John's Wort, Red sage, Golden seal, Chamomile, Clary sage, Geranium, Fenugreek, Goat's rue, Meadow sweet, Black horsebound, False unicorn root, Poke root, Wormwood, Barberry.

CHAPTER 5

Menopause

Case Presentations

- A fifty-eight-year-old widow and housewife with limited income lives with her unmarried sister. Her menstruation stopped at forty-two. Since then, she has had hot flashes, frequent irritability, mood swings, nervousness, and sleep disturbances. Her blood cholesterol and sugar are abnormal, and she is noncompliant with her prescription medication. Her diet is poor in minerals and vitamins.

- A fifty-one-year-old patient had increasingly irregular menses. Her last period was three months ago. She had been experiencing a sudden onset of profuse, embarrassing diaphoresis and the sensation of heat. She found her emotional state increasingly labile. She was irritated easily and had frequent mood swings.

Introduction

Menopause literally means "the end of monthly cycle," and signals the end of the fertile phase of a woman's life.

Even though menopause happens to occur between ages forty and fifty -five, the average age of a woman entering menopause is fifty-one (National Institute on Aging).

Menopause is a major issue in women's life after age forty when she ceases to menstruate and stops reproducing. It is caused by a decrease in the production of sex hormones by the ovaries. Symptoms include estrogen deficiency, hot flashes, irritability, vaginal dryness, and decrease in libido, osteoporosis, and genital atrophy. The atrophy also leads to dyspareunia (painful intercourse and vaginal bleeding).

During menopause a woman goes through psychological and hormonal changes In this phase of the developmental cycle hot flashes, nervousness, and disturbances in sleep patterns tend to occur. Women also experience depression, changes in self-image, social isolation, and identity crises.

In fact, humans are one of the few species in which females undergo menopause. It has been proposed that menopause increases a woman's overall reproductive success by allowing her to invest more time and resources in her existing offspring and/or their children rather than continuing to bear children into old age (Love 2003, 34 Menopause Symptoms).

Women's Health by WHO

"Women in Their Eighth Decade: What we know is that the average life span for women in America is now 81, and many of us can expect to live to 100. What are women in their eighth decade doing with their time? What are their hopes, dreams, expectations? What does "old" look like these days? We decide to find out

Biological and gender-related differences have given a shape to a man and a woman's health in society. The health of a woman or a girl, in many societies have disadvantage because of the discrimination rooted in that particular society based on sociocultural factors. Poverty yields a higher burden on women and girl's health because of involvement in unsafe cooking fuels (COPD), unequal power relationship between men and women, decrease education and paid employment opportunities, physical, sexual and emotional violence. Women and girls also face increased susceptibility to HIV/AIDS.

Physical Changes during Menopause

- atrophy in vagina, cervix, and uterus
- loss of vaginal muscle tone
- labia become less elastic
- loss of tone in muscles around bladder and rectum
- loss of tone and shape of breasts and redistribution of body fat and changes in body hair.

Drug Treatment

Hormone Replacement Therapy (HRT) is the most common drug treatment prescribed for symptoms of menopause. Estrogen is thought to help slow osteoporosis and reduce a woman's risk for heart attack and heart disease, two of the leading killers of postmenopausal women.

Risks of Using HRT

HRT possibly causes ovarian, uterine, and breast cancers. It has also been linked to autoimmune disorders such as lupus. Therefore, the best alternative should be to give a boost of estrogen-like compounds (phytoestrogens) in the diet.

In addition to HRT, tranquilizers, antidepressants, and sleeping pills are sometimes prescribed for helping menopausal symptoms. These drugs can cause their own side effects, which are discussed in the section of premenstrual syndrome.

Prevention and Decreasing the Severity of Symptoms Prior to Menopause

A good diet with nutritional supplements and stress management may help delay the onset of menopause. A positive lifestyle and regular exercise will improve calcium metabolism and help strengthen bones and avoid osteoporosis. A diet that includes protein, vitamin B complex, whole grains, meat, fish with lots of omega-3 and essential fatty acids, nuts, seeds, milk, and green leafy vegetables will help delay menopause.

Postmenopausal women should regularly consume about 1,000-1,200 mg of calcium, 500-1,000 IUs of vitamin D, and 500-800 mg of magnesium per day. Taking boron (2-3 mg) and fluoride (2-4 mg per day) will protect against osteoporosis.

Diet and Lifestyle Adjustments

- Avoid excessive intake of salt, chilies, sweets, spices, sour foods, and pickles.
- Decrease intake of caffeine and other stimulants, refined sugar, and cold drinks.

- Completely avoid junk foods and carbonated drinks.
- Increase intake of warm foods and drinks and add spices such as fennel and cumin.
- Go to bed early.
- Massage daily with almond or olive oil.
- Perform meditation, yoga, and regular exercise (walking).

Many Ayurvedic herbs can help to strengthen and rejuvenate the female reproductive system, regulate hormones, improve mental clarity, and calm emotions. These include aloe vera gel, shatavari (asparagus), saffron, kapikachhu (*Mucuna pruriens*), and ashwagandha (*Withania somnifera*). Jeshthamadh (licorice) can help correct hormonal imbalances. Beetroot juice is also helpful.

Herbal Remedies

Wild Yam Cream

Wild yam cream is made of wild yam by adding ¾ cup of oil: apricot, avocado or almond. You can use a combination of all three oils or try other oils.

- 1/3 cup cocoa butter or coconut oil
- 1 tsp. lanolin (lanolin can be replaced by shea or mango butter if you are allergic to lanolin)
- 1/2 oz. beeswax (beeswax can be omitted by using more coconut oil combined with shea butter or mango butter
- 2/3 cup distilled water

Wild Yam Root (decoction)

- wild yam extract (2 tsp.)
- aloe vera (1/3 cup)
- vitamin E (as desired)

Melt the oil and allow it to cool. Mix with 2/3 cup distilled water in a blender and slowly add the oils after it cools. The mixture will look like frosting. Pour into a jar and have fun. (Clark2000)

Herbs that Alleviate Menopausal Symptoms

- chaste tree berry (2 parts)
- wild yam (2 parts)
- black cohosh (1 part)
- goldenseal (1 part)
- life root (1 part)
- oats (1 part)
- St. John's wort (1 part)

In case of heart palpitations, high blood pressure, or tension, use motherwort in place of St. John's wort. In case of anxiety or depression, skullcap or valerian can be added.

Other Herbs that help with menopause are Dong quai (two capsules, 2-3 times daily)

Ginseng, cohosh, licorice root, unicorn root, sarsaparilla root.

Kelp tablets support the thyroid.

Cleavers and pokeroot herbs help clear the lymphatics. Astringents are used to help aid the healing of infected tissues, especially in the cases of mucus discharge. Types of astringents used include American cranesbill, beth root, false unicorn root, life root, oak, and periwinkle.

Use this Mixture for Best Results:

- American cranesbill (2 parts)
- beth root (2 parts)
- echinacea (2 parts)
- periwinkle (2 parts)
- cleavers (1 part)

Echinacea, garlic, and wild indigo are antimicrobials. Diets should be rich in vitamins, minerals, vitamin C, and garlic.

Ayurvedic View and Menopause

In Ayurveda, menopause is linked with normal aging, which is a *vata* (air)-dominated stage of life. Therefore, the symptoms and suffering in menopause experienced by a person are similar to the symptoms seen when the *vata dosha* is upset. *Vata*-type menopausal symptoms include depression, anxiety, restlessness, irritability, and insomnia. Menopause may also disturb the other two *dosha*s. Women with *pitta* (fire)-type symptoms are often angry and suffer from hot flashes. *Kapha* (water)-type symptoms include listlessness, weight gain, and feelings of mental and physical heaviness.

Ayurvedic treatment of menopause focuses on strengthening and rejuvenating the reproductive system. Diet and lifestyle adjustments are advised to maintain health and wellness. Herbal preparations and general tonics are prescribed to regulate hormones, calm mental stress, improve physical well-being, boost immunity, and provide rejuvenation.

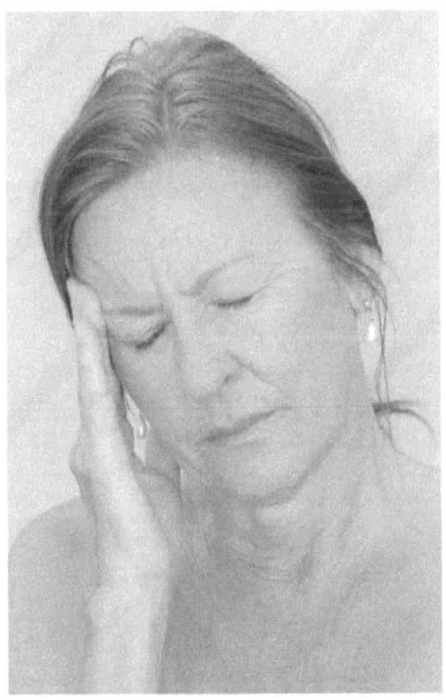

Ayurvedic Herbs for Treating Menopause

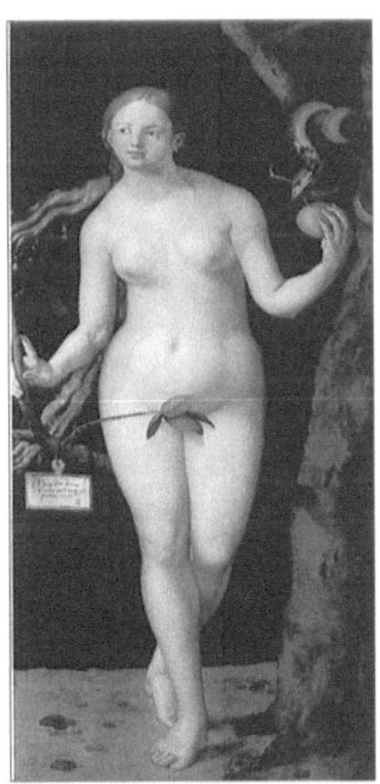

Ashwagandha

Known as Indian ginseng or winter chasteberry in English, Ashwagandha has been used for thousands of years in India to treat various symptoms of menopause, including insomnia, hot flashes, night sweats, depression, and moodiness.

Saffron

Saffron is a food additive with medicinal properties. It is widely used to help treat hot flashes and periods of irritability, annoyance, or anger. Those seeking nonprescription medications and relief for symptoms such as hot flashes and night sweats may find saffron quite effective. It's important to take saffron according to your own severity of symptoms. Follow instructions on supplements or advice offered by alternative medical or complementary medical practitioners.

Shatavari

Known in English as asparagus, it is commonly used by women to treat menopausal and postmenopausal symptoms, such as hot flashes and night sweats. The herb offers a natural estrogen source that helps provide relief and restoration of women's hormonal system imbalances. It is available as leaves, flowers, or roots.

CHAPTER 6

Impotence

Introduction

Approximately 43 percent of American women report dissatisfaction with their sexual function according to Laumann's study published in *JAMA* in 1999.

"Female Sexual Dysfunction: A Mind -Body Approach" by Cheryl B. Iglesias, MD, FACOG, explains the main reasons for this dissatisfaction: lack of interest in sex, arousal difficulties, problems with decreases in lubrication, pain during sex, anxiety about sexual performance, medications such as SSRI for depression, beta blockers for high blood pressure, oral contraceptive pills, lack of orgasm, hysterectomies, removal of ovaries, menopause, and ovarian cancer. Approximately 43 percent of American women are unhappy with their sex lives.

Impotency or sexual coldness is simply a permanent or temporary inability to have or maintain a satisfactory sexual relationship with one's partner.

Male or female impotence is noted almost in all cultures and depends on factors such as hormonal imbalance, disease, stress level, coping skills, depression, and socio-economic condition, especially in females. Women experience decreases in excitement and orgasm.

In my medical practice and in herbal practice, I have consulted many couples as well as individual female cases. Many women have come to my practice primarily because of sexual impotence or physical or mental stress. Whatever the cause, they are best managed with herbal products. After three to six weeks of treatment for sexual impotence, they have received satisfactory results from herbal remedies combined with psychological support, counseling, and education for coping skills to deal with interpersonal relationship and conflicts.

I have dealt with families who have been separated from loved ones because of immigration problems, single parents, separation, divorces, and infidelity. All these issues end up in depression, unsustainable and broken social lives, and seeking solutions for sexual impotence.

Many studies and writings are available in this regard. In order to solve this frustrated but treatable issue, we should focus on the causes of the problems. Open discussions with your partner and educate yourself to manage and maintain your sexuality. Don't be afraid to get professional help if needed. You are not alone, and you're worth it. Think about lifestyle changes; getting enough rest and exercise can improve your sex life.

As this is a common situation that occurs in all cultures, Eastern and Western authors have written about how to treat this reversible condition. Many of those who are suffering from sexual impotence feel depressed, have low self-esteem, and feel hopeless or helpless.

Kama Sutra

In order to improve your sex drive, consider changing your sexual routines. Candles, sexy clothing, music, *Kama Sutra*, exercise, sexual massage, and natural hormones can make it more pleasurable. Here, *Kama* means sensual desire or pleasure. Sutra is a thread or collection of verses.

Sex and romance have existed since the dawn of time. The *Kama Sutra* is one of the best-known books about love and lovemaking. It was translated from ancient Sanskrit to English by Sir Richard Burton in the mid-1800s. It is an ancient Hindu text known to mankind for centuries. The *Kama Sutra* is widely studied in this regard by Eastern and Western pioneers and writers. I will introduce very briefly, a part of this study, which is known to mankind for thousands of years. In this study, it indicates how to solve the impotency issues related to female sexual dysfunction.

The *Kama Sutra* is considered the primary Sanskrit work on human sexuality. It was written by Mallanaga Vatsyayana in the second century CE. Although Burton published it, this is the most widely known English translation. An Indian archaeologist,

Bhagvanla Indraji, performed the bulk of the translation with the assistance of a student, Shivaram Pashuram Bhide. This is one of the first systematic studies of human sexual behavior in world literature. It also documents the sociology of sex in India eighteen centuries ago.

The *Kama Shastra* is one of the three ancient Indian texts concerning the aims of life. In some texts *Kama Shastra* is the same as *Kama Sutra* . Two other Shasta are *artha* (material life), and *dharma* (religion, moral life). *Kama Sutra* should be understood within the context of *Artha Shasta* and *Dharma Shasta*. All three were written in Sanskrit in the seventh century BC.

The complete *Kama Sutra*, is translated by Alain Danielou in 1994, describes three kinds of activities that complete life's necessities:

"To assure its survival (existence), and its nourishment; reproduction according to forms of activity based on sexuality; and, lastly, to establish rules of behavior that allow different individuals to perform their roles within the framework of the species."

In human society, this is represented as three necessities and aims of life: material goods (*artha*) assure survival; erotic practice (*kama*) assures the transmission of life, and rules of behavior of a moral nature (*dharma*) assure the cohesion and duration of the species.

Kama is the third goal of human life that is fulfilled along with *dharma* and *artha*. *Kama* is defined as the enjoyment of worldly objects by the five senses—hearing, feeling, seeing, tasting, and smelling—assisted by the mind with the soul. The ingredient in this is a peculiar contact between the organ of sense, its object, and the consciousness of pleasure that arises from that contact. This whole process is called *Kama*. *Kama* is learned from the *Kama Sutra* (aphorisms on love) and from the practice of citizens (*Kama Sutra* 1883).

Ancient Indians believed that sex and survival were two of the most fundamental forces driving our continued existence. According to Alain Danielou, the *Kama Sutra* is an evolutionary work in that it promotes the cultivation of skills to become a well-rounded and well-evolved individual with healthy, intimate relationships with others. It is a picture of the art of living for the civilized and refined citizens, completing in the sphere of love, eroticism, and the pleasure of life.

The aim of the sixty- four arts of the *Kama Sutra* pertaining to wife of his own or other women represents the expression of love and intimacy by a man who is looked upon with love by his wife or the wives of others. In this art, cooking, bed making, coloring faces and body parts, sexual gestures, tattooing, dancing, flowering, putting on garments, jewelry, perfumes, and mental exercises such as puzzles, making figures, and images in clay are practiced to gain confidence in lovemaking. This art, if practiced by females, represents her to be a skillful, playful, understanding, refined, sexual, beautiful, and intelligent woman.

The *Kama Sutra* also expresses the male understanding of female nature and the importance of the relationship and intimacy. The ancient Indians showed great attention to the details of smell, light, music, food, drink, and touch before intercourse can begin.

Treatment

In order to better approach the female sexual dysfunction, we address it in medical and nonmedical types of managements.

Effective treatment for sexual dysfunction often requires addressing an underlying medical condition, such as a thyroid problem, diabetes, other hormonal conditions, depression, anxiety, or pelvic pain.

Antidepressants, antianxiety medication, and hormonal therapy such as estrogen in the forms of vaginal rings, creams, or tablets are available. Androgen therapy, which includes male hormones such as testosterone, is used to treat sex drive and libido. Tibolone, a synthetic steroid, is used in Europe and Australia. This drug was used to treat postmenopausal osteoporosis, which was accidently found to improve sexual function in women.

All of these treatments have potential side effects. The risk of hormone therapy depends on whether estrogen is given alone or in combination with progestin. It also depends on age and health issues, such as risk of heart or blood vessel disease and cancer. Antidepressants such as SSRI and antianxiety medications such as Benzodiazepines medication have been extensively studied and were mentioned in the section on PMS.

Tibolone is not recommended by the Food and Drug Administration for use in the United States because of increased risk of breast cancer and stroke in women. Research is needed before these agents might be recommended for treatment of female sexual dysfunction.

Nonmedical treatments include counseling, healthy habits, strengthening pelvic muscles by doing Kegel exercises, lubricants for treating dyspareunia and vaginal dryness. Vibrators or clitoral vacuum section devices are frequently used but have manifested unsatisfactory results.

Communicating your concerns with your body and its response to sexual activity is important for finding the real cause of sexual dysfunction. Women who are suffering from sexual dysfunction or who have sexual concerns often benefit from a combined treatment approach that addresses medical, relationship, and emotional issues.

Prevention

You cannot avoid all the risk factors for female sexual dysfunction, but you can:

- Follow the treatment if you have a medical condition.
- Ask your doctor to change the medication if the medication is the problem.
- Freely discuss your emotional stress with your care provider.
- Ask for relationship counseling if you have interpersonal conflicts.

Herbal Remedies

Female Potency Three X (www.naturagenics.com 2012)

A proprietary blend is available in capsule form, which contains more than twenty herbs in different concentrations. Examples are dong quai, damiana, oat straw, ginseng, black cohosh, red raspberry, extracts of adrenal, and ovary. I have suggested to many of my clients to take this preparation (two capsules twice daily for six to nine weeks) with satisfactory results. Psychological support and improving self-esteem are important factors to be combined with herbal remedies.

Herbal remedies such as oats, skullcap, and lime blossom may be indicated as nervine relaxants and tonics.

Ayurvedic Herbs

Shatavari (Asparagus)

This rejuvenative herb for *vata* and *pitta* promotes vitality and strength and is excellent for the female reproductive system, infertility, and sexual debility.

Shatavari plays a big role in female fertility and is the best-known female rejuvenator. It intensifies fertility by nourishing the ovum and organizes the womb for fertilization. The softening action of the herb decreases dryness on the vaginal wall. Shatavari helps ease premenstrual syndrome symptoms, controls blood loss during menstrual cycle, and controls ovulation. According to Ayurveda, shatavari is the best female reproductive system herb; it is often used for infertility, endangered miscarriage, leucorrhea, and menopausal problems.

Shatavari is given two teaspoons twice a day with milk or honey for its aphrodisiac property. Continue with this dose until satisfactory results are desired.

Shilajit

This Ayurvedic herb has been known for more than four thousand years for its balancing property for *vata*, *pitta*, and *kapha*. Shilajit reduces sexual debility in men and women.

Shilajit is popular in the preparation of Ayurvedic medicines and is considered one of the most significant elements in the Ayurvedic system of medicine. It helps control sex hormones for proper performance. It is also very beneficial in increasing an individual's sexual powers.

Ashwagandha (Withania Somnifera)

This Ayurvedic remedy balances *vata* and *kapha* constitution and promotes vitality and strength. Ashwagandha is traditionally known as Indian ginseng, and it has been used for centuries to deal with infertility and impotence in men and women.

This "adaptogen" herb promotes nerves and nerve function. It helps the body decrease stress, one of the most common causes of sexual problems. The roots of Ashwagandha are particularly beneficial in healing the disorders of reproductive systems of men and women.

Ashwagandha is easily digested when taken with honey, ginger, warm milk, meals, or hot water.

CHAPTER 7

Cancer

Case Presentations

- In September 1980, an eight-year-old boy was admitted with a diagnosis of leukemia to the pediatric ward of the Indira Gandhi's Children Hospital in Kabul. He was the tenth child in his family. He was very articulate and pleasant, but he was emaciated and weak because of the blood cancer. His father loved him so much and was a government employee, driving a school bus. He was ready to spend every last penny to get him treated and was even willing to sell his kidneys if needed to get more money for his son's chronic condition. The child was on chemotherapy, an expensive medicine at the time, but unfortunately the boy expired after forty-five days in spite of the best care.
- In December 1997, I was posted at Charity Hospital at Louisiana State University in New Orleans. I looked after and managed the HIV/AIDS patients who were suffering from severe complications, such as meningitis, and multi-organs involvement with the AIDS. They could not survive in spite of all the care given to them. I watched patients who were dying from encephalitis and cerebral involvement from AIDS complications. Their responses to the medications and nutritional adjustments were reported to be poor.

One of the major reasons that AIDS patients contract cancer very easily and cannot survive is the weakened immune system.

Definition

The National Cancer Institute defines cancer as "a term used for diseases in which abnormal cells divide without control and are able to invade other tissues and organs." Cancer cells can spread to other parts of the body through the blood and lymph systems.

There are more than a hundred types of cancers. Most cancers are named for the organ or type of cell in which they start; for example, cancer that begins in the colon is called colon cancer, cancer of the liver, called Hepatoma, etc.

Main Categories of Cancer

- Melanoma and Carcinoma begins in the skin or the tissues that lines the internal organs.
- Sarcoma is found inside the bone, cartilage, fat, muscle, blood vessels, or other connective tissues.
- Leukemia starts in blood-forming tissues, such as bone marrow.
- Lymphoma and myeloma begin in the cells of the immune system.
- Glioma, Astrocytoma, Glioblastoma multiforme, Medulloblastoma (central nervous system cancers) begin in the tissues of the brain and spinal cord.

One of the major types of cancer that female patients contract is invasive cervical cancer. This affects the cervix, the entrance from the vagina to the uterus. Almost all women who get cervical cancer also have human papilloma virus (HPV). HIV and HPV make cervical cancer grow faster.

Please refer to cancers of female genital organs in their particular section

Prevention

Herbs that support and repair the liver are important for fighting cancer. Abnormal cell growth is often a response to a continual irritant, such as cigarette smoking, the body's own waste products, waste products of disease -causing bacteria and other organisms, and environmental toxins such as pesticides. The liver produces enzymes that help the body break down and get rid of toxic waste. A decrease in liver function has been linked to an increased chance of developing cancer in different organs and takes its specific name.

What can you do to lower your chances of getting cancer?

Healthy Living and Early Detection

Always follow the recommendation of your primary health care provider for screening. Areas screened are the breast, cervix, and colon to detect any abnormality for primary prevention.

The seven dietary suggestions of the American Cancer Society (ACS) are important:

- Avoid obesity.
- Cut down on total fat intake.
- Eat more high-fiber foods, including whole grains, fruits, and vegetables.
- Eat cruciferous vegetables, such as cauliflower, broccoli, and cabbage.
- Eat foods that are rich in vitamins A and C.
- Lower alcohol consumption.
- Lower intake of salt-cured, smoked, or nitrite-containing foods.

All of these suggestions will help improve the immunity level of your body, which is important for fighting cancer.

Chickweed nourishes the glandular and lymphatic systems. Fresh tincture in a dose of forty drops (twice daily) is recommended for ovarian cysts, ovarian cancer, and other reproductive cysts (Weed 1989, 120).

Seaweed is another good ally against breast cancer, ovarian cancer, ovarian cysts, fibrocystic disease, menstrual irregularities, infertility, and menopausal problems (Weed 1989, 227).

CHAPTER 8

Diet, Nutrition, and Daily Vitamin Requirements

As a physician and an herbalist, I have consulted more than 900 female cases. In my seven years of herbal practice, I have noticed that a majority of them have no or minimum awareness about healthy nutrition and vitamin requirements. Many of my clients are concerned about the daily vitamin requirements, brands of vitamins, as well as the price of vitamins (out-of-pocket expenses not covered through third parties). I am obligated to shed some light on this issue to alleviate the confusion about what brand and what dose to take.

In general, a majority of clients ask about the correct daily dose of vitamins and wonder why there is such a great range in prices. A brand that contains 50 mg of vitamin B will probably cost more than one that has only 15 mg of vitamin B. As for the daily requirements of the vitamin, please refer to the table below (RDI) or read the print on the back of the bottles carefully in order to get the correct dose of the vitamins and minerals, including antioxidants.

The defining factor in terms of price is the fact that some very expensive supplements are hypoallergenic. These supplements are free of extra ingredients that could cause reactions in sensitive people. If you notice headaches, nausea, fatigue, palpitation, or loose bowel movements after taking the supplement, you may be reacting to an ingredient in that particular supplement. For example, most vitamin C is derived from corn, and many supplements are prepared in a yeast base. Some people react to these base ingredients. They should avoid taking these supplements. Hypoallergenic supplements are more expensive than others.

Another factor that affects the price of vitamins is that some are defined as "natural." This usually means that they cost more. I am introducing a chart for daily nutrition requirements written by my famous herbal instructor, Demetria Clark, MH. The charts cover

"

lifestyle changes, which indicate a move toward a vegetarian diet within the past decade.

A Simple Guide to Good Health for Moms and Newborns

(Wikipedia and Clark 2000)

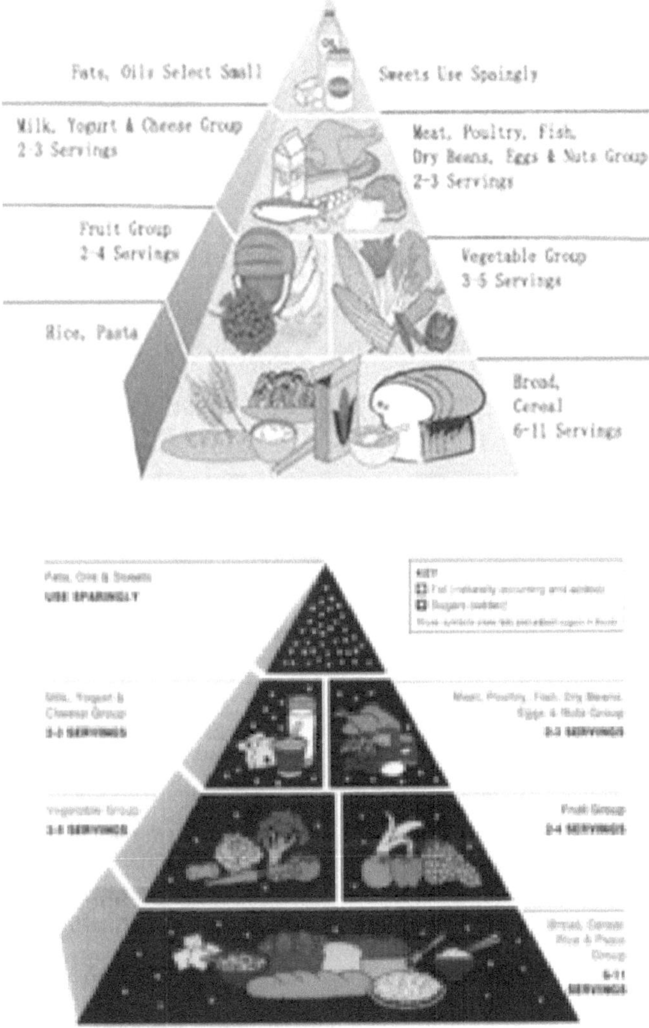

The USDA's updated food pyramid from 2005

The USDA food pyramid was created in 1992. It is divided into six horizontal sections containing depictions of foods from each section's food group. It was updated in 2005; colorful vertical wedges replaced the horizontal sections. It was renamed "My Pyramid" and was often displayed without the food images, creating a more abstract design.

In an effort to restructure food nutrition guidelines, the USDA rolled out its new "My Plate Program" in June 2011. It is divided into four slightly different- sized quadrants; fruits and vegetables take up half the space, and grains and protein make up the other half. The vegetables and grains portions are the largest of the four.

Step 1.

Eliminate Harmful Foods and Substances	Recommended Substitutes
White, brown, or natural sugars	Honey (raw only), maple syrup, stevia, anise, fennel. Stevia helps normalize blood sugar and is safer than most synthetic sweeteners.
Ordinary table salts with aluminum additives	Roasted sea-salt, miso salt (promotes beneficial bowel flora)
Margarine, Crisco, hydrogenated oils, oil-roasted foods (all are major factors to arthritis illness, heart disease and cancer)	Butter, ghee, mechanically pressed (cold or expeller pressed) vegetable oils (olive, sesame, and almond are best). Store in a cool, dark place. Do not cook with vegetable oil.
Fried foods in general	Add oil to foods after they have been cooked or steamed. Occasional sautéed onions or garlic are okay.

Avoid artificial colors, preservatives and/or synthetic ingredients. They are major contributing factors to ADD and ADHD in children.	
Ice creams (besides margarine) are the next worse substance for aggravating arthritis and TCM phlegm disharmonies.	
Most commercial dairy products, except for butter (pasteurized and homogenized dairy products in the United States are hard to digest and aggravate all types of TCM phlegm disharmonies).	Certain soy milks or rice milk (many show food sensitive), requires personal experience for substitute for milk
Avoid cold foods in general (Digestive enzymes do not work well if stomach fluids are too cool.)	Foods and drinks should be at least room temperature.
Commercial meats with additives that may include hormones, antibiotics, or formaldehyde	Organically grown meats (commercial lamb is often cleaner chemically than beef or chicken)
Shellfish and nonscaly fishes are highly susceptible to bacterial, parasitic, chemical, and heavy-metal contamination	Freshwater fish, scaly type sea fish from unpolluted waters. Inspect all fish for visible parasites; cook or bake thoroughly.
Hard-boiled or fried eggs	Soft-boiled or lightly poached eggs are easier to digest.
Commercial spices and herbs from many grocery stores are usually very old and are irradiated.	Nonirradiated spices and herbs. Fresh spices should smell fragrant. Obtain whole and grind just prior to use.

Foods canned in metal containers (acid foods may leach out metals from the container lining)	Foods canned in glass jars
Peanuts tend to be contaminated with harmful molds.	Almonds and filberts
Yogurt that is not completely soured is susceptible to unhealthy bacterial growth.	Completely cultured yogurt (may taste sour to be beneficial and promote healthy intestinal flora) with no additives and made with nonhomogenized milk
Flours of all types	Grind grains yourself just prior to use. Flour spoils quickly.
White flour lacks many of the nutrients that are present in wheat bran and germ. Most commercial flour is chemically bleached.	Whole grain flours, ground as needed
Baking powder containing aluminum	Baking powder containing calcium carbonate as the active ingredient. Too much baking powder may create alkaline intestinal pH; combine with yogurt to minimize interference with vitamin metabolism.
Wheat germ spoils rapidly.	Use whole wheat, which has not had its germ removed
Moldy or green potatoes	Fresh potatoes (cut off all dark, eroded areas before baking)
Tobacco, alcohol, drugs, coffee, tea, chocolate	Roasted chicory or barley; various nonmedicinal herbal teas

Step 2.

Eat high-quality foods	Watch for allergies, dairy and wheat intolerances, corn, eggs, nuts, seeds, and citrus fruits.
Basic food groups	**Recommended foods**
Meats, fish	There is no need for meat consumption in this society, but if you choose to eat meat, please use the highest ethical and conservation practices.
Nonhomogenized dairy products	Butter, cheese, milk, goat's milk, yogurt (sour). A majority of people benefit by eliminating all dairy products (except for butter, which is better than margarine).
Eggs	Eggs
Nuts and seeds	Walnuts, pecans, pine nuts, filberts, almond, and pumpkin seeds
Beans and bean products	Black turtle, red, kidney, garbanzo, pinto, aduki, mung, lentil, white navy; soy products (better for Asian people who possess the necessary enzymes to digest properly; includes soy milk and tofu)
Grains and starches	Wheat products (includes breads, pastry, pasta, triticale, semolina, couscous, gluten, seitan, spelt, kamut, tabouleh, and bulgar), rye, barley, oats, corn, millet, rice, buckwheat (toasted or untoasted) amaranth, quinoa, sweet potatoes, yams, potatoes, arrowroot powder

Vegetables	Starchy vegetables: sweet potatoes, yams, potatoes, pumpkins, winter squashes, peas, corn Root: carrots, beets, parsnips, salsify, turnips, and kohlrabi Greens: *Chenopodiaceous* family (eat sparingly because of oxalic acid content), beet greens, chard, spinach Crucifers: cabbage, Brussels sprouts, broccoli, cauliflower, kale, collards, and mustard greens Misc.: zucchini, summer squash, string beans, okra, eggplant, green peppers, and tomatoes
Fruits	Concentrated-sugar fruits: figs, dates, dried fruit in general, carob, and tamarind Misc.: bananas, pears, grapes, pineapples, plums, peaches, nectarines, cherries, strawberries, papayas, and apples Citrus: oranges, grapefruits, lemons, and limes
Seasonings	Culinary herbs and spices, miso, sea salt (roasted), soya bouillon or tamari, tamarind, cured green or black olives, onion, garlic, lemon, lime; apple cider vinegar; mechanically-pressed sesame, almond and extra-virgin olive oil; baking powder (nonaluminum variety); tomato sauce (jarred, no added sugar)

Based on Rocky Mountain Herbal Institute Teaching

Step 3. Follow Basic Rules for Food Combining

- Eat fruits and simple sugars like honey and maple syrup alone on an empty stomach. Wait several hours before eating anything else. (Exceptions: Apple, as well as using lime or lemon to season other food.) Fruits are absorbed quickly by the digestive tract, but combining them with other foods causes the sugars to ferment within the digestive tract, promoting unhealthy bacterial flora, yeasts, and parasite growth.
- Do not eat starches and meats in the same meal, unless you possess an "iron" stomach. (If you do this, it is best to eat meat first, followed by starches, since this optimizes the pH necessary for each in the stomach and duodenum). For example, combine rice, beans, and vegetables. Do not combine meat and potatoes or rice.
- Eat beans, grains, or potatoes in a 1:3 ratio (less beans more grains). This optimizes the proportions of various amino acids needed to rebuild body proteins and is a ratio followed throughout the world, especially in cultures with low animal protein intake.
- Soak beans overnight in warm water with one tablespoon of apple cider vinegar. After eight hours of soaking, drain off vinegar and soak in water. Add cold water to about three times the level of the beans. Add a few teaspoon of thyme and cook twice as long as standard recommendations; i.e., two hours for lentils, eight hours for garbanzos.
- Do not eat within several hours of going to bed. At night, the body likes to be empty of food.

(Clark 2000)

Step 4. Cooking Procedures

Stainless steel (magnetic type) and Pyrex cookware are the best. Do not use Teflon or aluminum pots and utensils. Aluminum reacts with food, especially acidic foods, and elemental aluminum will react with digestive enzymes to deactivate them. Chronic aluminum poisoning is also implicated in Alzheimer's disease and other degenerative CNS diseases.

The best cooking methods are baking, steaming, and simmering. High temperatures damage food; avoid frying, stir-frying, or barbecuing food.

Avoid using microwave ovens, which are health hazards because of radiation exposure and the risk of cancer.

Step 5. Use Moderate Amounts of Spices and Seasonings

Try using miso or lemon instead of salt. Use asafetida, cayenne, chili peppers, black pepper, and vanilla.

Essential oil extracts include orange, lemon, bitter almond, citronella, and vanilla. Use stevia rebaudiana as an herbal sweetening agent if sugar intake must be limited for health reasons.

The recommended daily intake (RDI) of vitamins and nutrients, which is regulated by the Food and Drug Administration (FDA) is considered to be sufficient to meet the requirements of 97-98 percent of healthy individuals in every demographic in the Unites States.

The RDI is used to determine the daily value (DV) of foods, which is printed on nutrition facts labels in the United States and Canada. The following tables list the DVs based on a calorific intake of 2,000 calories for adults and children four or more years of age.

Total Fat	65 g
Saturated Fatty Acids	20 g
Cholesterol	300 mg
Sodium	2400 mg
Potassium	3500 mg
Total Carbohydrate	300 g
Dietary Fiber	25 g
Protein	50 g

Nutrient and Vitamin Deficiency and their Effects on the Body

Nutrient	Deficiency	RDI	Highest RDA (Recommended Daily Allowance)
Vitamin A	Reduced cellular immunity, slow tissue healing, increased infection rate, decreased defense to mucous membranes	900 mcg	900 mcg
Vitamin C	Decrease phagocyte function, reduced cellular protection, and slow wound healing	60 mg	90 mg
Calcium	Osteoporosis, bone fractures, skin, nail and heart abnormalities	1000 mg	1300 mg
Iron	Decreased cellular immunity and neutrophil activity	18 mg	18 mg
Vitamin D	Impaired calcium absorption	400 IU (10 mcg)	600 IU
Vitamin E	Decrease antibody production, lower cell-membrane integrity	30 IU	15 mg (33 IU of synthetic)
Vitamin K	Bleeding impairment	80 mcg	120 mcg
Thiamin (B1)	Enlarged heart, decrease in alcoholics	1.5 mg	1.2 mg

Riboflavin (B2)	Bloodshot eyes, sore tongue & lips, infection in the mouth & throat, sensitivity to light, chapped lips	1.7 mg	1.3 mg
Niacin (B3)	Coated tongue, dizziness, forgetfulness, insomnia, irritability, dementia, diarrhea	20 mg	16 mg
Vitamin B6	Lessened cellular immunity, slow energy metabolism	2 mg	1.7 mg
Folate (Folic Acid)	Reduced blood cell production, increased cervical cancer, neural tube defects in pregnant women	400 mcg	400 mcg
Vitamin B12	Decreased lymphocyte proliferation and PMN bactericidal activity	100 mcg	100 mcg
Biotin	Hair thinning and impairment	300 mcg	30 mcg
Pantothenic Acid (B5)	Lowered humoral immunity, increased irritation and stress	10 mg	5 mg

Phosphorus	Weak bones or teeth, joint pain & stiffness, decrease energy, lack of appetite	1000 mg	1250 mg
Iodine	Impairs thyroid metabolism	150 mcg	150 mcg
Magnesium	Nervous system abnormalities, important for bone strength	400 mg	420 mg
Zinc	Decreases T and B cell function and thymic hormones; increased infection rates and slow healing	15 mg	11 mg
Selenium	With vitamin E deficiency, antibody response is lowered	70 mcg	55 mcg
Copper	Lowered resistance to infection	2 mg	900 mcg
Manganese	Neurological symptoms (ataxia), hearing loss, fainting, diabetes, infertility	2 mg	2.3 mg
Chromium	Anxiety, fatigue, diabetes	120 mcg	356 mcg
Molybdenum	Problems related to liver (jaundice), heart (tachycardia)	75 mcg	45 mcg
Chloride	Heavy sweating, congestive heart failure	3400 mg	2300 mg

Summary

"Childless women are more likely to die prematurely. Women are fortunate and blessed with the ability to bring forth young creatures and nourish them with milk. A Danish study of more than 20,000 couples determined that becoming a parent greatly decreases your risk of premature death" (NBC's Brian Williams). Dec, 06, 2012.

The WHO defines health as "a state of complete physical, mental, psychological, emotional, spiritual, and social well-being, and not merely the absence of disease or infirmity."

Reproductive health, therefore, implies that people are able to have a responsible, satisfying, and safe sex life—and that they have the capability to reproduce and the freedom to decide if, when, and how often to do so.

Women face unique health issues. From puberty though menopause, they deal with hormonal fluctuation, monthly menses, ovulation, pregnancy, and lactation. They may suffer from premenstrual or menopausal tensions, including irritability, mood swings, breast tenderness, uterine discomfort, headaches, insomnia, nausea, and occasional vomiting. Some may deal with uterine fibroids, breast and ovarian cysts, yeast infections, vaginal dryness, cervical dysplasia, and incontinence.

Breast cancer is another major issue; as many as 44,000 women die from breast cancer each year. Many more suffer from chemotherapy, disfiguring surgeries, and radiation, damage to self-image and sexuality, or painful death.

Aging women are warned to guard against osteoporosis and heart disease (375,000 American women die from heart disease each year). Coronary risk factors such as obesity, diabetes, and hypertension are all on the rise among women.

A woman's birthright is good health, strength, and vitality. There are simple ways to nourish the body, ease daily discomforts, and prevent life-threatening diseases.

As an herbalist and nutritional counselor, I'd like to share my views and values about healthy lifestyle that I have adapted. With blessings for the knowledge I have received from well-known gurus and advisors, I will share my thoughts about nutrition and medicinal herbs in the hope that all women can enjoy healthy lives.

CHAPTER 9

Materia Medicas

Information on different herbs (*Materia Medicas*-www.healthy.net) comes from David Hoffmann, Demetria Clark, Rosemary Gladstar, and Gail Faith Edwards.

Black Cohosh (Cimicifuga Racemosa)

It is found in wooded areas in the United States and Canada. Roots and rhizomes are dried. It has a powerful action as relaxant in cases of painful or delayed menstruation. It helps with ovarian cramps. It can be used during labor to aid uterine activity. It should be avoided during pregnancy. It is available in powder, capsule, and liquid forms.

Angelica Sinensis and Dong Quai

For thousands of years, healers around the world have used these herbs to regulate menstruation, relieve cramping and menstrual distress, promote healthy blood circulation, and balance hormones.

These two herbs improve the sexual and libido problems, alleviate constipation, and improve sleep. Both provide rich stores of vitamins A, B (especially B12), and E. Those B vitamins, in combination with lots of niacin, magnesium, and calcium help strengthen nerves and relieve tension. Vitamin E in this herb helps to keep skin, internal organs, and tissues, especially those of the bladder and vagina, lubricated, moist, and flexible. Angelica's high iron content nourishes and builds blood, prevents anemia, and increases vital energy.

Supplies of phytosterols (hormonal precursors found in plants), glycosides, saponins, and flavonoids support the body's production of all important hormones, stabilizing emotional swings, and easing hot flashes, irritability, and hormone-related headaches. Both herbs are rich in coumarin derivatives, which promote antispasmodic and vasodilatory effects. They are useful for relieving muscle tension and painful menstrual cramps.

Angelica can regulate menstruation after coming off the pill and relieve symptoms of menstrual discomfort.

Chasteberry (Vitex Agnus-Castus)
(Clark 2000; Hoffmann 1996)

The fruit of the dark berry is used for medicinal purpose. It increases the progesterone-estrogen ratio by decreasing FSH release. It acts as a uterine tonic through stimulating and normalizing the pituitary gland, especially its progesterone function.

The greatest use of chasteberry is normalizing the activity of female sex hormones; it is used for dysmenorrheal, premenstrual syndrome, irregular menstruation (especially if accompanied with endometriosis), and fibroid cysts. It is also used for acne in teenagers, treating conditions for withdrawal from birth control pills, and regulating the ovarian cycle if a woman wants to get pregnant. It also increases milk production for mothers who want to breastfeed their babies.

According to herbalist David Hoffman, the most important use of *Vitex* in England is for treating menopausal symptoms such as hot flashes, irritability, and vaginal dryness. In 1930, Dr. Gerhard Madus researched an extract of dried fruit named Agnolyt and developed a medicine. After that, all the studies on *Vitex* have been done with his patented preparation. He found it to have a "strong corpus-luteum effect, which increases progesterone."

Vitex has no reported side effects. Infusion, capsule, and tincture forms are available.

Wild Yam (Dioscorea Villosa)

Wild yam is an herb used to ease dysmenorrheal, ovarian, and uterine pains. It is used as decoction, tincture, and creams.

The root and bulb of the plant are used in preparing extract or liquid forms. It is promoted as a natural alternative to estrogen therapy after it has been processed in a laboratory. It is used as estrogen replacement therapy for vaginal dryness in older women, premenstrual syndrome (PMS), menstrual cramps, increasing energy and sexual drive, and breast enlargement.

Some women apply wild yam creams to the skin to reduce menopausal symptoms such as hot flashes.

It is also used in early stages of rheumatoid arthritis when there is intense inflammation. It has strong antispasmodic and anti-inflammatory activities.

The Role of Shepherd's Purse in Female Reproductive Health
(Hoffmann 1996; Clark 2000; Ellingwood)

Shepherd's Purse (Capsella Bursa-Pastoris)

Shepherd's purse can be found almost anywhere in the world. The plant has a white flower throughout the year.

Shepherd's purse is used to stop heavy menstrual bleeding, particularly from the uterus. It has been used to treat postpartum hemorrhage. It is also used for premenstrual problems and menstrual cramps.

It is effective in treating chronic uterine bleeding disorders, including uterine bleeding due to the presence of a fibroid tumor. It contracts the uterine muscles and stops any hemorrhaging from the genital tract or the gastrointestinal tract such as with bleeding ulcers.

Do not use shepherd's purse in pregnancy or while breastfeeding internally or externally. Skin application can cause the uterus to contract, causing menstruation and miscarriage.

To make an infusion, use two teaspoons of dried herb with boiling water; drink every two or three hours during and just before period. It can also be used as a tincture (1-2 ml, three times a day).

Red Clover (Trifolium Pretense)

Red clover is especially rich in minerals, including calcium and magnesium, and has abundant amounts of protein, vitamin B complex, and vitamin C.

The National Cancer Institute has found anti-tumor properties in this herb . . . Antioxidant known by tocopherol, a form of vitamin E has been shown to help prevent breast cancer in animals. Red clover is prescribed today for the treatment of breast, ovarian, and lymphatic cancers. (Edwards, 2011).

In order to treat these diseases, it is best to drink two to four cups of red clover blossom daily. You can also massage red clover-infused oil into affected area daily; use the fresh flower or a poultice over the area.

Phytosterol-rich red clover blossoms have been used to balance hormones and promote the production of estrogen. Menopausal women reduce hot flashes and night sweats with red clover usage. Gail faith Edwards said, "Red Clover is the number one fertility enhancing herb."

It regulates the menstrual cycle, but it thins the blood better. It should not be used for heavy bleeding because it contains salicylates and coumarins. It is safe for pregnant women and nursing mothers. It will help produce abundant amounts of breast milk.

Red Raspberry (Rubus Ideaeus)
(Hoffmann 1996)

Red raspberry has abundant amounts of calcium, phosphorus, potassium, and vitamins B, C, and E. It has an alkaloid, fragrine, which strengthens the entire pelvic region, especially the uterus and ovaries, helping the uterus in preparation for childbirth.

Its high calcium content is beneficial for the nervous system and relieves pain in childbearing and labor. If nausea or morning sickness is an issue, it is better to drink the infusion.

In order to have a successful, full -term pregnancy and to promote conception, leaves are excellent fertility-enhancing herbs. They are often combined with red raspberry blossoms to help control heavy menstrual bleeding.

Red raspberry is used in infusion and tincture forms.

Yarrow (Achillea Millefolium)

Yarrow helps slow excessive menstrual bleeding and shrinks fibroids in premenopausal women. Yarrow balances hormones, especially progesterone, in the body. A yarrow tincture helps slow menstrual bleeding (ten drops every two hours until bleeding slows).

Gail Faith Edwards recommends taking twenty drops of yarrow tincture or one cup of infusion twice daily to help shrink fibroids. Yarrow also works as a disinfectant and has antibacterial properties that fight urinary tract infections and incontinence.

Lady's Mantle (Alchemilla Vulgaris)
(Clark 2000; Hoffmann 1996; Edwards 2001)

This herb has been widely used throughout Europe for its astringent, emmenagogue, and anti-inflammatory qualities. Its uterine tonic effect regulates menstruation, reduces pains associated with periods, and eliminates excessive bleeding. As an emmenagogue, it stimulates proper menstrual flow. Because it contains salicylic acid, it alleviates menstrual cramps and uterine discomfort. It is used to shrink fibroids and relieves breast discomfort by dealing with lumpy breast tissue.

The flowers and leaves are used as a poultice on breast tissue, increasing tone and firmness. It also helps with vaginitis, genital sores, vulvitis, herpes, and perineal tears by drying up excessive discharge. In order to regulate menstruation, Gail Faith Edwards recommends ten drops of lady's mantle three times a day for a week before menstruation is due.

Lady's mantle is used as infusion of the dried herb or in decoction form. Besides poultice, it is available in tea and tincture form. In low doses, there has been no toxicity noticed.

The dose is adjusted according to the severity of symptoms. It is better to take in three divided doses.

Partridge Berry and Squaw Vine (Mitchella Repens)
(Hoffmann 1996)

Native North Americans discovered squaw vine. It is among the best remedies for preparing the uterus and whole body for childbirth. It should be taken for some weeks before the child is due, ensuring a safe and protected birth. It may be used to relieve painful periods (dysmenorrheal).

Partridge berry is recommended for sore nipples. It helps to relieve painful menstruation, cramps, ease labor and labor pain. It also helps women for childbirth.

To prepare for childbirth, it may be used with raspberry leaves. For dysmenorrheal, it may be combined with cramp bark and pasque flower.

It can be used as an infusion in boiling water or in tincture forms.

Blue Cohosh (Caulophylum Thalictroides)
(Clark 2000)

Found in the United States, the roots and rhizome are collected in the autumn and are rich in natural chemicals.

Blue Cohosh is an excellent uterine tonic. It is used at any time during pregnancy if there is a threat of miscarriage. Because of its antispasmodic action, it will ease false labor pains and dysmenorrheal. Its use before birth will help ensure an easy delivery. In all these cases, it is a safe herb to use. As an emmenagogue, it can be used to bring on delayed or suppressed menstruation.

Special indications in its use are: Endometriosis, ovaritis, dysmenorrheal, urethritis, vaginitis, restlessness during pregnancy, menopausal pains and discomfort. In other words it decreases the inflammation and spasmodic activity in all above conditions.

To strengthen the uterus, it may be used with false unicorn, motherwort, and/or yarrow. Combine it with skullcap and/or black cohosh to increase its antispasmodic effects.

It is used in decoction and tincture forms.

Motherwort (Leonurus Cardiac)

Motherwort is beneficial for female health. It is used in those who suffer from PMS, delayed menstruation, and emotional imbalance associated with premenstrual conditions. It is also useful for anxiety related to thyroid hyperactivity, heart palpitations, or nervous origins.

I have used motherwort combined with hawthorn, nettle, skullcap, and bugleweed (twenty to thirty drops, three times a day) for a few cases of hyperthyroid with palpitation and anxiety with satisfactory result. It is recognized as a tonic herb for the uterus in regulating the menstrual cycle and relieving menstrual cramping.

For girls starting puberty, this herb is beneficial. According to Gail Faith Edwards, a dosage of ten drops mixed with water a few times a week will help regulate menstruation.

Shatavari (Asparagus)

Shatavari is native to India, belongs to the family of asparagus. It has nourishing, soothing, and cooling properties. Therefore, it helps with many conditions in which the body and mind are overheated, depleted, or out of balance.

Researchers have paid more attention to Shatavari's immune-modulating properties. Studies show that Shatavari strengthens the immune system by enhancing the functions of macrophages (the immune cells responsible for digesting potentially destructive organisms and cancer cells).

Its botanical name is Asparagus racemosus. In Sanskrit, *shita* means cold, and *vari* or *virya* means energy. The name translates to "she who possesses a hundred husbands," referring to the herb's rejuvenative and energizing effects upon the female reproductive organs.

Besides treating female reproductive organs, it also treats gastrointestinal disorders such as hyperacidity, stomach ulcers, dysentery, and bronchial infections for men and women. It soothes dry, irritated membranes of the upper respiratory tract and reduces dryness of the vaginal wall.

Properties

- nutritive tonic, rejuvenative
- aphrodisiac
- galactagogue
- laxative
- antispasmodic
- antacid
- diuretic
- antitumor
- demulcent

Medicinal Use

It is the most important herb in Ayurvedic medicine for women. It nourishes and cleanses the blood. Internally, it is used for infertility, nourishing the ovum, loss of libido, threatened miscarriage, and menopausal problems. It is a good supplier of female hormones.

Shatavari also helps improve PMS symptoms by relieving pain and controlling blood loss during menstruation. Shatavari also supports the normal production of breast milk.

In males, it is used in sexual debility, impotence, spermatorrhea, and the inflammation of sexual organs.

REFERENCES

Alan R, Gaby, ed. 2006 *The Duke Encyclopedia of New Medicine: Conventional and Alternative Medicine for All Ages.* London; New York -"American Pregnancy Association, "Promoting Pregnancy Wellness," http://www.nlm.nih.gov/medlineplus/pregnancyloss.html.

Archives, August 31, 2011 Breastfeeding Basics

Berst, Truman. 2013. Alternative Health and Herb Remedies. Albany,

Brian William's Report, Dec, 2012, A Danish Study "Childless Women More Likely to Die Prematurely."

Brown, Harry. 2005. Review from *Netlines. British Medical Journal* 331:1345.

Burton, Richard. 1883. Kama Sutra.

Center for Disease Control and Prevention 2011

Clark, Demetria. 2000. *Heart of Herbs, Volume I and Volume II.*

Coffey, Lisa Marie. 2004. *What's Your Dosha, Baby? Discover the Vedic Way for Compatibility in Life and Love.*

Daniélou, Alain. 1993. *The Complete Kama Sutra: The First Unabridged Modern Translation of the Classic Indian Text.*

Darroch, JE, et al. 2001. "Difference at Teenage Pregnancy Rates among Five Developed Countries: The Role of Sexual Activity and Contraceptive Use." *Family Planning Perspectives* 33(6):244-50.

Edwards, Gail. 2001. *Herbs for Women's Health.*

Giller, Robert M., and Kathy Matthews. 1994. *Natural Prescriptions.*

Gladstar, Rosemary. 2001. *Family Herbal: A Guide to Living Life with Energy, Health, and Vitality.*

http://www.henriettesherbal.com herbalsafari.com. "Health and Ayurveda" hivinsite.ucsf.edu/InSite, pg-kb-05-02-03

Hoffmann, David. 1996. *The Complete Illustrated Holistic Herbal.*

Holly Lynn Padove, California College of Ayurveda, July 2003.

http://www.bbc.co.uk/news/health-20613131

http://www.nlm.nih.gov/medlineplus/miscarriage.html.

Lad, Vasant. 2001. *Textbook of Ayurveda, Volume One: Fundamental Principles*. Ayurvedic Press.

Lad, Vasant. 2007. *Textbook of Ayurveda, Volume Two: A Complete Guide to Clinical Assessment*. Ayurvedic Press.

Love Susan, MD, *Menopause and Hormone*, New York, Three Rivers Press, 2003

Mayo Clinic 2010 *Book of Alternative Medicine, Second Edition.*

Mayo Foundation for Medical Education and Research (MFMER). 1998-2012.

"Miscarriage" Medicine Plus Accessed May 24, 2012 Rodriguez 2012, http://natural-fertility-info.com.

"Royal College of obstetricians and Gynaecologist. Accessed September 2012."

http://www.rcog.org.uk/about-us,

www.obgyn.uab.edu/medicalstudents

www.naturagenics.com 2012, Chino.

Weed, Susun S 1989 *Wise Woman Herbal, Healing Wise.*

White and Foster. 2000. *The Herbal Drugstore.*

GLOSSARY

abstain: To prohibit ones self from doing something, or stay away from something that brings joy or misery to one self.

accumulation: the process or action of gathering something aching joints: Pain where two bones or ligaments join in a human or animal body.

acquaint: To become familiar with

acquire: To gain/ to earn

adaptogen: An herb that increases the body's ability to adapt to stress and changing situations. Example: Ginseng

addictive: Causing an addiction of some sort

adhering: To give support or maintain loyalty

aggravate: To make worse, more serious, or more severe

agni: Fire/heat

AIDS: Acquired immune deficiency syndrome.

ailment: A bodily disorder or chronic disease

alleviate: To get rid of a pain

amenorrhea: Absence of menses (periods)

amnesia: Forgetfulness

androgens: A male sex hormone

anesthetic injections: Relating to or resembling anesthesia, cause numbness so that no pain is felt

antibacterial: Kills and prevents the growth of all bacteria.

antibiotic: Kills or prevents the growth of living organisms, especially viral and bacterial infections.

anticancer: Prevents the formation and growth of cancer cells.

anticipation: Realizing in advance or ahead of time, before something happens

anti-inflammatory: Any drug or medication that helps reduce or prevent inflammation

antioxidant: Prevents oxidation. In humans, oxidation is in part

responsible for signs of aging and formation of free radicals
   antiretroviral drugs: Drugs that kill the HIV/AIDS virus
antispasmodic: Eases or prevents spasms, especially in the muscles.
aphrodisiac: Increases sexual desire and ability, usually by
providing optimum nourishment to the endocrine system
ART: Antiretroviral therapy.
arthritis: Painful inflammation of the joint
astringent: A medical substance that contracts the tissues or canals of
   the body, thereby reducing secretions and diminishing swelling.
atrophic vaginities: Is an inflammation of the vagina (and the outer
   urinary tract) due to the thinning and shrinking of the tissues, as
   well as decreased lubrication.
attributable: A quality or characteristic inherent in or ascribed to
   someone or something
augmenting: To make greater as in size, extent, or quantity
ayurveda: The knowledge of life or living, known to mankind for
   thousands of years.
benign: Medicine that has little or no dangerous effect; harmless
   blissful:Extreme happiness
bloating: Make or become swollen due to gas or excess fluid
   boullion: A broth made by stewing meat, fish, or vegetables in
   watercandidiasis: An infection of candida that causes oral of
   vaginal thrush
cavity: An empty space or gap in a solid object or human body part
central nervous system: The complex combination of nerve tissues
   that control most or all the activities in the body, plays a major
   part in the brain and spinal cord.
cervical effacement: thining of the cervix
colitis: inflammation in the colon
colostrum: rich in antibodies, first glands to appear after labor
combat:Fight againstcomplementary feeding: when breast feeding is
   not enough, a mother must start to give her child taste of food.
   Ex: banana, and blended vegetable, or brand name foods such as
   Gerbert.
conjunctivitis: The inflammation in the mucous like membrane in
   the eye
consume: To inhale or take in

contraception: Artificial methods or techniques to prevent pregnancy, such as birth control pills

carcinogenic: Substance that can cause cancer.

child bearing period: The time where a female is physically ready to have a child

chiropractic: Using manual power to remove pressure from the spine, nervous system allowing them to function more efficiently.

chlamydia: A very small parasitic bacteria that depends on other cells to reproduce

chemotherapy: The treating of a disease by using chemical substances

clitoris: A small sensitive and erectile part of the female genitals at the anterior end of the vulva.

colic: Pain in any part of the body such as abdomen caused by accumulation of the wind (vayu).

constipation: Difficulty in evacuation of the bowel.

compliance compliant: To follow the suggestions about medicines, diet, and supplements.

conventional medicine (allopathic medicine): Deals with prescription drugs.

cortical areas: Any region that surrounds the cerebral cortex; which is in charge for conciousness

corpus luteum: A hormone-secreting structure that developes in an ovary after ovum is discharged and it usually disappeares, unless a pregnancy is started

coumarin: Vanilla scented compound found in many plants

chromosomal: Of or in relation with chromosomes

complications: Hardships or a difficult situation

condiments: Spices

cramps: Painful, involuntary contraction of a muscle

craved: The sudden urge of wanting of something, more commen when it comes to foods

cruciferous: Denoting plants of the cabbage family

culinary herbs: Herbs that are commonly used in food, cooking, or in kitchen

cultivation: The act of preparing or loosening and breaking up (tilling) of the soil. The soil is prepared for plant growth by destroying weeds.

cystitis: bladder infection causing infrequent and painful urination.
debilitating cramps: Mild discomfort that may cause vomiting, bloating and abnormal bowels movements
delusional thinking: A false idea caused by mental illness, a false psychotic belief outside the self that is not true. An evil spirit.
demulcent: The relieve of inflammation
devastating: A sad or tragic occurrence
diaphoresis: A side effect of a prescribed medication, usually involve sweating
diindolylmethane (DIM): A natural remedy for balancing estrogen
dilemma: Trouble, problem
disillusioned: Disappointment that something is not as good, true, or valuable as it seemed. Having lost faith or trust in something.
disinfectant: A chemical liquid that kills or destroys bacteria
distress: Extreme sorrow or anxiety
diuretics: Drugs which cause increase in urination
dominant: Something that over powers something else
dormant: A period in an organism's life cycle when growth, development, and physical activity are temporarily stopped.
douching: A cleansing solution used to wash a body part or cavity (as the vagina).
dosha: Ayurvedic concept with three main principles
dysmenorrhea: Severe pain accompanying monthly period.
emmenagogue: Promoting the normal flow of menstrual blood. A substance that stimulates the menstrual flow
dysparenia: Painful intercourse
elements: A fundamental substance that is made of any one kind of atoms, which constitutes a matter. Any of the four substances water, air, fire, and earth believed to be a physical universe.
eliminate: To get rid of, erase
encephalitis: The inflammation of the brain, caused by infection, mostly virus etiology.
endocrine system: The hormonal secretions that are formed in the gland or a part of gland called the endocrine system, help to integrate and control the bodily metabolic activities such as thyroid, pituitary, adrenal gland, and more.

endometriosis: A condition that happens from the appearance of endometrial tissue outside the uterus and causes pelvic pain.

enhance: To increase or make greater in quality

entity: Anything that has independent existence

eroticism: The state of sexual arousal. A quality that causes sexual feeling.

estrogen: Any of a group of steroid hormones that promote the development and maintenance of female characteristics of the body

etiology: A branch of medical science concerns with the causes and origin of diseases.

excruciating: Very painful

expelling: To kick out or revoke from a program, group, or membership

explore: To try to find undiscovered places, animals, or things in general

facilitate: To make something easy

feminization: Development of female characteristics

fertilization: Involving the fusion of male and female gametes to form a zygote.

fetus: offspring not yet born

fibroids: The beginning of a tumor, usually around the uterus wall

flatulence: Gas made in the stomach and intestine as food breaks down into energy

flavonoids: Any member of a class of biological oxygen-containig aromatic antioxidants compounds but no nitrogen group found in many plants.

foul odor: A very bad smell

freckles: Dot like speck that occur on a persons face

free radicals: Uncharged molecule produced in the body by natural biological processes or introduced from an outside source (tobacco, pollutants, and toxins) that can damage cells, proteins, DNA by altering their chemical structure.

frothy: Yellow / green discharge of the vagina with a foul odor usually to some sort of infection

FSH: Follicile-stimulating Hormone

galactagogue: Increases the quality and quantity of breast milk.

genital tract: The lower urinary and reproductive part of a woman

gestation: The process of a fetus being carried between the time of conception and birth

glands: A group of cells in the body that selectively removes materials from the blood or alters and secretes them for further use in the body or for elimination from the body-compare endocrine gland, exocrine gland.

glycosides: A compound made of simple sugars and a separate compound that replaces the hydroxyl group in a normal sugar molecule.

GnRH: Gonadotropin releasing hormone

groin region: The inner section in between the thigh with the trunk, often include the external genitals

grounded: Describes a person who is mentally and emotionally stable, sensible and has a good understanding of what is really important in life.

gyn: Abbreviated form of Gynecology

hallucinations: The experiencing of something that may seem to be there but in reality is not. Examples: Visual-seeing things, Auditory-hears things.

harmony: Coordination, equilibrium, balance in life, surrounding.

healing: The process of becoming better or healthy again

hepatic herbs: Strengthens and nourishes the liver and its functions.

hepatoma: Liver cancer.

herbal medicine: Herbs, herbal materials, herbal preparations, and herbal products that contain parts of plants, other plant materials, or combinations as active ingredients

holistic health: Concerns with wholes or complete system (body, mind and soul) rather than with parts of the body.

holistic approach:. To deal with the whole body (body, mind, and soul).

hot flashes: A sudden feeling of heat, usually a common symptom of menopause

HPV: Human Papilloma Virus

human papillomavirus (HPV): Virus that causes cervical cancer.

hypnotic: Induces or encourages sleep and sleepiness.

hypoallergenic: Does not cause any allergic reaction

hypothalamus: Region of the brain that controls both automatic nervous system, body temperature, hunger, thirst, emotional activity and sleep

hysterectomy: Removal of uterus

immunity: The ability of an organism to fight infection and disease

impotence: Low in sexual desire and orgasm

impurities: The quality or condition of not being pure or clean

incidence: New cases of a disease

incontinence: Unable to control urine, feces, etc.

infant: A child less than 12 month old

infertile: Unable to conceive

infertility: Inability to conceive

infirmity: Physical or mental weakness

inflammation: A physical ability where a part of the body becomes red, swollen, hot, and even painful in some occasions

innate: Natural

insomnia: Deprived from sleep

intercourse: Sexual act

intervention: A process applied to prevent harm or improve functioning such as application of medication to prevent arrhythmias.

invades: To occupy

irritation: The state where a person becomes annoyed, impatient, or angry

kama sutra: An ancient Indian Hindu text on the study of human sexual behavior.

kapha: One of the three principles of Ayurveda manifested by accumulation of the toxins and phlegm in the body

labia minora: Smaller inner folds of the vulva

labia majora: The larger outer fold of the vulva

labile: Easily changed

lactation: The process of milk by the mammary glands

laser therapy: Medication, treatment of therapy where laser is the primary source for the cure

lectins: Specific carbohydrate binding protein

lesions: Injury

LH: Lutenizing hormone

lash out: To aggressively attack someone in some sort of physically or verbally

leucorrhea: Yellow like discharge of the vagina

leukemia: A malignant progressive disease where excessive leukocytes are formed

libido: sexual desire

life spans: How long something lives; human, animals, or even plants

listlessness: Feeling lack of interest or energy

lumps: Compact mass of any substance

lymphatics: A veinlike vessel that carries out lymph in the body

lymphocytes: A type of small leukocyte or a part of white blood cells that causes a cellular immunity.

lymph nodes: Where lymph is filtered and lymphocytes are made

macrophage: A large phagocytic cell found in tissues when it's not in motion and in white blood cells when it's moving

mansa: Tissue

malformation: Something that doesn't come to a full formation or is incomplete

mastitis: Inflammation in mammary glands found in the breast or under usually caused by a bacterial infecton on the nipple or teat

menarche: First time a menstruation happens

menorrhagia: An excess loss of blood occurring during menstruation.

meditation: A technique of relaxation

meningitis: Inflammation of the maninges

mesh: Material made of wire or thread, such as a fishnet

metastases: The development of secondary malignant growths, after the first appearance of cancer

metabolism: A chemical process that happens in living organisms to stay alive

miracle: A surprising and welcomed event, not scientific law therefore believed to be the work of a devine agency

miscarriage: The expulsion of a fetus; can be volunteerly or the result of an accident

monks: A member of a religious community of men

morbidity: The relative or relating to a specific disease

mortality: The state of being subjected to death

metrorrhagia: Cause bleeding in the middle of the menstrual cycle.

mucus: A slippery secretion that is rich in mucins and is produced by mucous membranes.

narcotics: An illegal drug that affect the mood or behavior of a person

nitrous oxide: A colorless gas with a sweet smell, used for anesthesia

nourishment: Food or other substance needed for growth, health, and good condition of a living thing; human, animal, or plant

nurture: To care for

obstacle: Something that may get in the way of a goal, desire, or wish

obscure: Uncertain to figure out

offender: A person who commits an illegal action

ojas: Vigor, strength, immunity

optimal: Best or most favorable

optimistic: To have a positive attitude

optimize: To make the best out of a situation

orgasm:A climax of sexual excitement

osteoporosis: Medical condition where the bones become fragile from loss of tissue, hormonal change, or lack of vitamin D

ovum: Mature female reproductive cell

ovulation: The release of an ovum from the ovaries, normally half way through the menstrual cycle

oxalic acid: A poisonous crystalline acid that has a sour tast, found in bleach and cleaning products

oxytocic herb: A polypeptide hormone produced by the posterior lobe of the pituitary gland

panchkarma: Five fold detoxification treatment

passion: A strong sexual or romantic feeling for some one. A strong desire for or devotion to some activity, object, or concept.

passionate: Showing or caused by strong feeling

pathogenic: Having the ability to cause disease.

pelvis:The wide curved bones between the spine and the legs.

perabens: Compounds that are used as preservatives in pharmaceutical products and food

perineal area: Area between the vulva and anus in a woman

phagocytic: A group of cells such as white blood cells that consumes and engulfs foreign material as microorganisms and debris.

phlegm: Sticky mucus secretions

phytosterols: Group of steroid alcohols found in small quantities in vegitable oils and corn oil

phytoestrogens: Estrogen that occurs naturally in legumes

pitta: One of the three principles of Ayurveda by accumulation of heat and toxins in the body.

pituitary glands: Major endocrine gland located in the skull and produces various internal secretions for conducting basic body functions.

placenta: An organ inside the mother's uterus, nourishes the unborn baby, and is pushed out of the mother after the baby is born.

PMS: Premenstrual Syndrome

poison ivy: North American plant, cashew family, contains an irritant oil within the leaves which may cause an itch

polymenorrhea: Excessive bleeding during menstruation or not during period.

polymorph nuclear (PMN): White blood cells.

post mastectomies: The after effect of mastectomy

poultice: Soft, moist mass of material usually made of a plant or flour, used to relieve soreness and inflammation

prakriti: Constitution, refers to a person's general health. The prakriti is believed to be unchanged over a person's lifetime.

prana: Vital life energy

premalignant lesions: Tissue in which cancer is more likely to occur

progesterone: A steroid hormone released by the corpus luteum that helps prepare for pregnancy

prolactin: A hormone that helps the production of milk after chid birth

prolapse: The occurrence of an upward or downward slip of an organ

proliferation: A fast increase in numbers

prominent: Important or famous

promiscuous: Having many sexual relationships

prone: More likely to

PTSD: A psychiatric disorder that can happen after witnessing a life-threatening event, abbreviated for Post-traumatic Stress disorder

puberty: The period where teenagers reach sexual matuarity

pungent: Having a strong smell or a sharp and bitter taste.

rasa: Taste

rasa rasayan: Get rid of toxic elements from cancerous cells.

rakta: Blood

RDI=recommended daily intake

recognize: the ability to identify

region: A section of

rejuvenative: Restores youthful qualities, increases stamina, and encourages joie de vivre.

reluctant: Unwilling

replicating: To repeat or make a copy some thing exactly same.

restore: To fix

resume: To get back to what was happening

retention: To keep

reveal: To express

rhizomes: A continuously growing underground stem

romantic: Feeling of excitement and mystery mainly associated with love

salicylates: A salt of salicylic acid

salicylic acid: A bitter compound found in some plants, used as fungicide

saponins: A toxic glycosides found in plants as soapwort or soapbark characterized by producing a soapy lather.

scabies: Contagious skin disease identified by itching and small red spots

screening tests: Strategy used to identify a disease.

sedative: To cause sleep

sexual development: Another term for puberty

sexual tonic: Strengthens and improves the functioning of the internal and external sexual organs, not an aphrodisiac.

Sexual transmitted disease: Infections that you can get from having sex with someone who has an infection

shifting dullness: A sign, elicited on physical examination for ascites

sinusitis: Inflammation of the sinus

spermatogenesis: The production of mature sperm

spermatorrhea: Excessive and involuntary ejaculation

spermicides: A substance that kills sperm

standardized: To make or to convert to a standard condition such as herbal products processed and maintained in standard condition with quality control and other issues.

stimulant: Increases or speeds up functional activity, often pushes one past normal limits.

strives: To make an effort to reach a goal

subjective findings: The expression of the symptoms of a disease by the patient

substances: A specific kind of thing or substance with the same properties

substantial: An importance in growth or size

symptoms: Physical or mental condition that is often a warning or sign of a disease

tamari: A naturally fermented soy sauce

tampons: A soft plug material that is inserted into the vagina to absorb menstrual blood flow, often used in place of pads

teenage: The period between the ages of 13 and 19

tejas: intelliegence

tendency: The inclination to a specific characteristic or kind of behavior

tics & twitches: An unvoluntary movement of the muscles, most commonly on the face

traditional medicine (TM): Knowledge, skills, and practices based on the theories, beliefs, and experiences indigenous to different cultures, used in the maintenance of health and in the prevention, diagnosis, improvement, or treatment of physical and mental illnesses. In some countries, it is referred to as "alternative" or "complementary" medicine (CAM).

tridosha: Three dosha or three principles of Ayurveda expressed as Vata, Pitta, kapha

trigger: A release or offset of a mechanism

trochomoniasis: An intracellular human pathogen.

trimester: Three month

urination: The discharge of urine

utilize: To use something

vaginitis: Inflammation of the vagina

vata: One of the three principles of Ayurveda by accumulation of air and gas in the body.

vedas: the most ancient Hindu writing, written in Sanskrit, contains hymns, philosophy and guidance

vital organs: Tissues that are joined in structural unit to serve a comman function

vitality: The state of being strong and having energy or being active

vulva: The external genital of a female

warts: Caused by a virus, small hard, growth of the skin

womb: The uterus

yeast infection: Infection caused by fungus

yoga: A Hindu spiritual practice of disapline, includes breath control. meditation, and specific body positions to help relax

yogis: A person who is proficient in yoga

yukti vyapashraya: To get rid of dead cells in the patient's body.

ABOUT THE AUTHOR

The Author Chela Ram is familiar with both treatment systems (Alternative and Conventional).He speaks English, Persian, Pashtu, Urdu, Hindi, and Spanish.

About 80 percent of his regular clients are Latinos.

Belonging to a family who loved to grow and nurture flowers and herbal plants, he has developed a pure nourishing and healing nature that has helped him in learning more about herbal medicine.

He has earned a master's degree in herbal health. After five years of herbal practice—and consulting about 1,250 clients with different health ailments—he took the American Herbalists Guild test, and was honored to become a registered herbalist, RH (AHG) in the United States.

The Author has consulted 900 female cases from a total of 1575 cases in past seven years with herbal practice. He has an extended medical experience with different health issues in Asia and United Staes in past 35 years.

This experience has enabled Chela Ram to adopt a strong belief system with herbal cure. The author wants to address in this book, how the female health can be managed in an affordable and safer way by preparing herbal teas, tinctures, and decoctions at home and change their life style with a little guidance about the herbs.

The author has extended educational background in Pediatrics, Psychology, Mental Disorders, Public Health, and Herbalism. The author has worked with children, geriatrics, inmates, individuals with drugs and alcohol addiction, depression, anxiety disorders, menopausal and sexual transmitted diseases which will direct and give the best decision making skills to someone who comes for health related assistance.

ABOUT THE BOOK

This is a comprehensive guide to the female reproductive system with Ayurvedic and Western herbalism. The ailments of female genitalia and herbal cures focus mainly on three age ranges (adolescent, childbearing and menopausal). Readers will easily adopt and learn about the genital ailments by reading the introductory outlines and will choose the treatment options that best fit their needs.

Several of the approximately nine hundred female cases consulted during the author's herbal practice, are presented in this book as an example. Herbal treatments and prescription drugs are outlined in this book. It also includes cures with Ayurvedic and Western herbalism, and many testimonials are described in detail.

INDEX

hypoallergenic 100, 134
hypothyroidism 55
hysterectomies 88
hysterectomy 35, 48, 50, 135

I

ibuprofen 26, 44
immune-modulating properties 124
immune system 19, 22, 24, 47, 97,
    98, 124
immunity xxvii, 18, 22, 24, 25, 53, 84,
    99, 109, 110, 135, 136, 137
impaired immune systems 12
Impairs thyroid metabolism 111
impotence xxvii, xxxiii, 88, 89, 96,
    125, 135
Impotence xxxiii, 88
impurities 5, 135
incarcerated 21
incontinence 112, 119, 135
increases milk production 75, 115
Indian ginseng 86, 96
infant 3, 74, 75, 76, 135
infertile 38, 67, 135
infertility 37, 38, 40, 55, 62, 64, 65, 66,
    67, 68, 94, 95, 96, 99, 111, 125, 135
Infertility 20, 66
inflammation of sexual organs 125
inflammations 7
inguinal nodes 17
insomnia xxvii, 26, 28, 38, 69, 84, 86,
    110, 112, 135
intercourse 11, 13, 14, 20, 21, 22, 37, 47,
    54, 60, 69, 78, 92, 132, 135
interpersonal relationship 88
intervention 35, 135
intravenous drugs 12
invasive cervical cancer 98
ivy reactions 7

J

Jeshthamadh licorice 82

K

Kama Shastra 90
Kama Sutra 89, 90, 91, 92, 127
Kapha 8, 42, 84
Kapha Dosha 8
kapikachhu 82
Kegel exercises 93
Kelp 83
Kutki 53

L

labia majora 3, 135
labia minora 3, 135
Labor 72
lactation 3, 4, 112, 135
lactobacillus 15, 16
Lady's Mantle 50, 120
lash out 7, 136
laxative 125
lesions 6, 12, 17, 18, 22, 66, 135, 138
leucorrhea 95, 136
leukemia 97, 136
libido 25, 26, 65, 78, 92, 114, 125, 136
licorice 18, 19, 28, 29, 30, 31, 82, 83
Licorice 31
licorice root 18, 28, 83
life forces
    doshas 6
life root 83
lifestyle changes xxviii, 35, 89, 101
Lily flowers 57
lime blossom 94
Lisinopril 50
lumps 13, 49, 50, 136
lumpy breast tissue 120
lupus 81
luteinizing hormone
    LH 3, 64
luteinizing hormones 49
lycopene 45
lymphatic 24, 49, 56, 68, 99, 117
lymphatic channels 49
lymphatic system 49, 56, 68

lymph nodes 17, 22, 136
lymphocyte proliferation 110
lymphocytes 22, 136
lysine 18

M

Maca 57
Macafem 36
maca root 57
macrophages 124
magnesium deficiency 25
male fern 19
malformation 61, 136
mansa 53, 136
marital conflicts xxxiii
massage 24, 35, 89, 118
mastitis 50, 136
Materia Medica xxxiv, 43, 50
Materia Medicas 113
maturation 4
meadowsweet 72
Meditation and Massage 24
melatonin 49
menarche 4, 136
meningitis 22, 97, 136
Menopausal 83, 118
menopausal symptoms 81, 84, 115, 116
menopause 2, 7, 9, 12, 47, 51, 78, 79, 80, 81, 83, 84, 86, 88, 112, 134
Menopause xxxiii, 8, 35, 36, 38, 78, 79, 80, 81, 84, 85, 128
menstrual bleeding 9, 33, 40, 117, 119
menstrual cramps 27, 34, 37, 38, 114, 116, 117, 120
menstrual cycles 9, 47
menstrual irregularities xxxii, 11, 33, 99
Menstrual Irregularities 33
metabolisms 7
metabolizer 56
metastases 50, 136
metronidazole 15
Miconazole 15
midwife cocktail 73

migraine headaches 27
milk thistle 45, 65
Milk thistle 41, 45, 58
milk thistle seed 58
miracle 32, 136
miscarriage 16, 38, 42, 43, 58, 74, 95, 117, 122, 125, 128, 136
Miscarriage 42, 128
miscarriages 6, 40, 42, 43
monks 5, 136
monounsaturated 52, 60
mood swings 15, 23, 25, 62, 69, 78, 112
MOPH
    Ministry of Public Health xxi
Morning Sickness 71
morphine 50
motherwort 29, 30, 83, 122, 123
Motherwort 31, 123
mucous membranes 109, 137
mucus 42, 83, 137
multiple sex partners 12
mushrooms 16, 49, 52, 61
My Pyramid 102
myrrh 18, 19

N

narcotics 72, 137
natural xxvii, 5, 7, 8, 16, 36, 38, 41, 50, 54, 55, 56, 58, 59, 74, 87, 89, 100, 102, 116, 122, 128, 132, 133
natural antiprostaglandin 50
natural hormones 36, 89
Natural progesterone cream 28, 56
nervine 94
nervousness 6, 25, 78, 79
Nervous system abnormalities 111
nettle 27, 57, 123
nettles 43, 77
Nettles 31
neural tube defects 110
neutrophil activity 109
Nimba 53
nitrous oxide 72, 137
nonhomogenized milk 104